DISASTER
PREVENTION,
PLANNING
AND
LIMITATION

DISASTER PREVENTION, PLANNING AND LIMITATION

Proceedings of the
First Conference
University of Bradford
12-13 September 1989

Editors

A. Z. Keller and H. C. Wilson

DPLU
University of Bradford

THE BRITISH LIBRARY

ISBN 0-946655-35-9
© 1990 The British Library and
Disaster Prevention and Limitation Unit, (DPLU),
University of Bradford

British Library Cataloguing in Publication Data

The Cataloguing in Publication Data for this book is
available from the British Library

Produced by Technical Communications (Publishing) Ltd.,
Letchworth, Hertfordshire, England.

FOREWORD

During 1988 the Science Reference and Information Service of the British Library, in developing its strategic plan for the next 5 years, made a strong commitment to provide literature and information services on the environment. We recognised that this issue was now an international problem on which our substantial resources could provide invaluable backup for research workers and scientists in the United Kingdom and other countries. As a visible indicator of our commitment to the environmental issues of today, we produced a major bibliography on Acid Rain during 1988, and this was followed in early 1989 by a volume of collected papers written by European scientists on Acid Deposition. In recent months, titles have been published on the United Kingdom Green Belt, the impact of the Channel Tunnel, and titles expected during 1990 include literature guides on Hazardous Waste, further volumes of collected papers on Acid Deposition and guides on Transport and the Environment and the Costs and Economics of Waste Recycling.

This volume of collected papers on Disaster Prevention and Limitation Planning is therefore one of an increasing number of British Library documents which disseminate information on environmental problems and related issues. There is considerable public concern in both this country and abroad over what might be called man-made disasters and their aftermath. The 16 papers in this conference proceedings address a number of issues concerned with such disasters, examining the role of the emergency services and the problems created by man-made disasters and their impact on the human and natural environment. Also, a number of papers address the possibilities of disaster prevention and examine the possibilities of prediction of disasters and the application of safety and reliability methodology.

Through our association with the British Library's Document Supply Centre at Boston Spa in Yorkshire, our close links with a number of major libraries and information services in this country who cover a wide spectrum of scientific and technical subject matter, and our access to an international range of environmental abstracts services and databases, we can provide a major support service to all those carrying out research into disaster prevention planning and limitation. The readers of this conference proceedings will find that virtually all the references to periodical papers, report literature and other material are available from either the British Library or the associated libraries working specifically in the field of disaster prevention.

A Gomersall
Director
The British Library
Science Reference and Information Service
25 Southampton Buildings
London WC2A 1AW

Telephone: 01 323 7485

CONTENTS

THE WORK OF THE DPLU
1988/1989

The initial aims of the DPLU were to establish a working unit within the University of Bradford which would carry out research related to prevention and mitigation of consequences of disasters.It would also disseminate information relating to disasters and act as a general forum.

The DPLU has an Advisory Committee of eighteen members drawn from the Emergency Services,Industry,European Commission,Health and Safety Executive,Local Authorities and Delft University.

Activities of the DPLU During the Period 1988/1989.

Activities of the unit in the past year were as follows

1).Research and Publications.

Researchers within the unit have devised a scale which classifies the severity of a disaster in relation to the number of fatalities that have occurred.Further work will be directed at extending this scale to incorporate environmental,financial and other factors.Work also continued on the development of risk indices for public water supplies.

Some of the principal publications by the unit are appended at the end of this report.

2).The First Disaster Prevention and Limitation Conference.

During September 1989 the unit held an Inaugural Conference at the University of Bradford which was attended by 140 delegates.The Conference was held over two days and topics covered the fields of disaster prevention,pre and post disaster planning and management.Delegates came mainly from Local Authorities,Emergency Services and industry.

The Conference attracted national interest and was extensively reported by the media.

Arising from the Conference the need for working parties to be established was identified.Areas to be covered by these working parties include the establishment up of a computerised database for use by the Local Authorities and Emergency Services.The need for a further database to include information with regard to the management of actual incidents was identified.Many of the smaller Local Authorities stressed a need for computer simulated and table top exercises to assist them in their emergency planning role.

A Committee has been formed to undertake the planning of the Second Disaster Prevention and Limitation Conference which again will be held at the University of Bradford on the 11th and 12th September 1990.

3).Other Activities.

A previously assembled bibliographical listing of relevant publications relating to disasters will be updated and issued.This work will be carried out under the supervision of the University Library staff..

The Institution of Chemical Engineers have invited the DPLU to run a short course in collaboration with them.The aim of this course is to increase the awareness of the smaller chemical companies with regard to the potential for disaster arising from such sites.The proposed date for this course is 26/27 March 1990.

One of the Merseyside Authorities has requested that the unit be involved in the evaluation of their preparedness should an incident occur within its boundaries.

A five day module on disaster prevention and limitation has now been introduced into the European Masters Course on Safety and Reliability.This module will be organised jointly by the University of Bradford and the University of Delft.

Research papers on the work of the Unit have been presented at an Invitationary Seminar on Disaster Management at Middlesex Polytechnic and at the Western Regional Conference of the American Society for Quality Control held in Salt LAke City.

Members of the Unit have been invited by a major publishing house to produce a text on the topic of transient water pollution incidents and their potential effects on the public.

Close contacts have been developed with other organisations involved with disaster prevention and limitation and requests for assistance or advice have progressively increased over the last year.

THE BRADFORD DISASTER SCALE

Dr. A. Z. Keller,
Chairman of Postgraduate Studies,
Department of Industrial Technology,
University of Bradford

INTRODUCTION

Hillsborough,Locherbie,Herald of Free Enterprise,Clapham Junction are some of the more notable disasters that have occurred within Europe over the past few years.The Armenian earthquake,Bangladeshi floods and the Sudanese famine and subsequent cholera outbreak have all claimed numerous lives outside Europe.

These are all referred to as "disasters".

The Oxford English Dictionary(1) states that the word "disaster" comes from the 16th century French word "desastre".The definition given by the O.E.D.is as follows;

>"Anything that befalls of ruinous or distressing nature;a sudden or great misfortune,mishap or misadventure;a calamity"

We suggest that a "disaster" may be defined as

>"an event which afflicts a community the consequences of which are beyond the immediate financial,material or emotional resources of the community. "

For the purpose of the present study a disaster is defined "when ten or more fatalities result from one event over a relatively short period of time."

The disaster may involve a man-made event such as a rail or air crash or may be a natural occurrence such as a violent storm.There may also be a combination of man-made and natural events.e.g.a combination of an inadequate dam design and abnormal weather conditions resulting in the collapse of the dam.

The event may be localised in geography and time as in the Clapham Common rail crash or it may involve a large part of a nation such as the Ethiopian famine,or the Bangladeshi floods.

Because of relatively large financial and material resources of the Western World many after affects and suffering can be ameliorated.

However,despite significant assistance from the developed
nations,mainly because of the relative underlying poverty of
the country,the effects of the Ethiopian famine are still felt
today.

What factors are required to enable a community to cope with a
disaster?
An organised infrastructure involving emergency
services,medical care and some form of after-care are obvious
candidates.Less obvious factors are community spirit,degree of
acceptance of events and initial quality of life within the
community.

Within the Western World the emergency services are generally
highly developed ,the same can be said to hold for the
transport infrastructure.This allows a fast response to a
disaster and rapid transfer of the injured from the site to
the medical care centres.These facilities are seldom available
in the under-developed nations,and,if available,are seldom
able to cope properly with the situation as it develops.

The under-developed nations,however, tend to have a more
localised well developed community spirit which is sometimes
lacking within the more developed nations.
The threshold of acceptance of events in the developed
nations is lower than in the under-developed nations.A mother
in Ethiopia may expect to lose many of her children before
they reach their teenage years.This is not a situation that a
mother in Europe or the U.S.A. would tolerate.This does not
imply,however, that the Ethiopian mother's grief is any less
real or intense.

To compare disasters of like occurrence or similar fatality
rate requires that each disaster be quantified numerically
before such a comparison is made.The Bradford Disaster Scale
and Classification System sets out a methodology for the
quantification of disasters based on the number of fatalities
involved.This allows disasters arising from differing sources
and causes to be compared.

In the analysis which is now presented information with
regard to disasters prior to 1976 have been extracted from
Nash (2),and after 1975 from The Times.

PRIMARY CLASSIFICATION OF DISASTERS.

Disasters can be categorised into three classes,

a).Natural Disasters.

Natural disasters generally are beyond the ability of man to produce,influence or prevent,e.g.earthquakes,volcanic eruptions,cyclones etc..The scale of loss of life from natural disasters can range from a few individuals to several million. Examples are,the Armenian earthquake,Krakatoa and the many cyclones and hurricanes that afflict the Philippine Islands.

b).Man-Made Disasters.

These are the disasters that are of anthropogenic origin.Examples are rail and air crashes,mining and marine disasters,large scale deaths due to the action of fires or explosions.The associated loss of life due to this type of disaster seldom exceeds several hundred.
Examples of these are, Clapham Common,Herald of Free Enterprise,Piper Alpha.

c).Hybrid Disasters.

These arise from a concatenation of anthropogenic(man-made) and natural events.
Man and his associated activities can produce natural disasters that would not otherwise occur,or significantly aggravate the effects of a natural disaster.
Examples of these are;the spread of disease from a community within which the disease is endemic to a community with no natural immunity to that disease,such as,the introduction of European influenza to the Eskimoes;the wholesale destruction of the Himalayan rain forests and consequent reduction in evapotranspiration which has intensified the annual flood occurrence of Pakistan and Bangladesh;the large scale deaths in the early 1950's due to smog production in London and other major U.K.cities.The loss of life due to this type of disaster can be, and usually is,extremely large.

For the purposes of this paper disasters have been classified by type into 17 categories and in Table 1 these have been referenced by the degree of man's involvement.

TABLE 1

DISASTER CLASSIFICATION
and PREDOMINANT AGENT

Disaster Type	Natural	Man-Made	Hybrid
Avalanche/Rockfall	Yes	No	Yes
Landslide/Mudslide	Yes	Yes	Yes
Air Transport	No	Yes	Yes
Climatic	Yes	No	?
Drought	Yes	Yes	Yes
Famine	Yes	Yes	Yes
Epidemic	Yes	No	Yes
Plague	Yes	Yes	Yes
Earthquake	Yes	No	No
Fire	Yes	Yes	Yes
Explosion	No	Yes	Yes
Flooding	Yes	No	Yes
Marine Transport	No	Yes	Yes
Mining	No	Yes	Yes
Rail Transport	No	Yes	Yes
Volcanic Activity	Yes	No	No
Miscellaneous	No	Yes	Yes

For the purposes of the analysis certain of the categories above have been grouped together as they can present difficulties in classifying source data.The differentiation between fire and explosion or avalanche and landslide is problematical and consequently fire and explosion will be referred to throughout as fire and avalanche and landslide as avalanche.

Drought,famine,plague and epidemic similarly present classification problems.For example,the large scale loss of life in Ethiopia is reportedly due to famine but the underlying cause of the famine is a long term drought within the region.The majority of fatalities within the region was not due to famine but to epidemics and plagues which readily spread through the under-nourished populace.Consequently for this presentation these four sources are referred to together as D.F.E.P.(Drought,Famine,Epidemic,Plague)

The query against Hybrid Climatic disasters is that although the activities of man can influence climatic effects no disaster that can be completely classified this way has yet been identified.

THE SCALE OF THE PROBLEM.

Within the last century over 400,000 people have lost their lives in disasters that have occurred within Europe.
31,500 of these fatalities occurred within the United Kingdom.

The total number of incidents occurring within Europe during this period is 1140,and within this period 270 incidents have occurred within the U.K.,i.e. 24% of the European total.
However,disasters occurring within the U.K. in general were smaller in scale.

Figure 1 gives a frequency curve for the 10,ten year intervals of the last century ,1888-1988,for disastrous occurrences within Europe.There was a gradual increase for the first eighty years followed by a much more rapid increase over the last two ten year periods.Figure 2 gives the corresponding figures for the U.K..The plot here is more erratic but shows a decline over the first 50 years to 1936 after which there has been a general increase especially over the ten year period covering 1978-1988.

Figure 3 gives the total fatalities for these ten year periods for Europe.Apart from two major peak periods,1908-1918 and 1918-1928, the number of fatalities are reasonably consistent through out.The 1908-1918 peak was due to the Italian earthquake in Sicily in 1918 and later on in that year,again in Italy,a massive flood.The 1918-1928 peak was due to a series of avalanches in the Italian Alps which caused many fatalities amongst troops stationed in that area.

 The 1908-1918 peak in Figure 4 for the U.K.was due to a large number of civil marine incidents that occurred throughout that period whereas the peak period 1948-1958 was due to the large number of deaths that resulted from the high level of smog in the winter of 1952.

Frequency of Disasters by Type

    The annual frequency of disasters that have occurred over the past one hundred years by type are shown in Figures 5 and 6 for Europe and the U.K. respectively.The major difference between the two sets of results is that the U.K.suffers fewer natural disasters than Europe.The frequency of man-made and hybrid disasters is greater than that of purely natural disasters.Air,marine and rail being the most frequent type of disaster incident.

    Although man-made and hybrid disasters are more frequent than natural disasters the number of lives lost in Europe in the latter is far greater.(Figure 7---1898-1908 and 1908-1918 results are reduced by a factor of ten for both

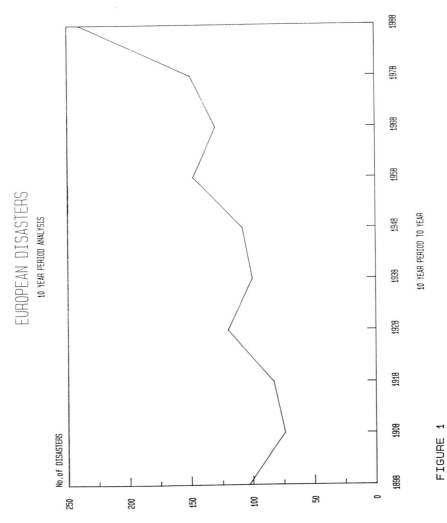

EUROPEAN DISASTERS

10 YEAR PERIOD ANALYSIS

No. of DISASTERS

10 YEAR PERIOD TO YEAR

FIGURE 1

8

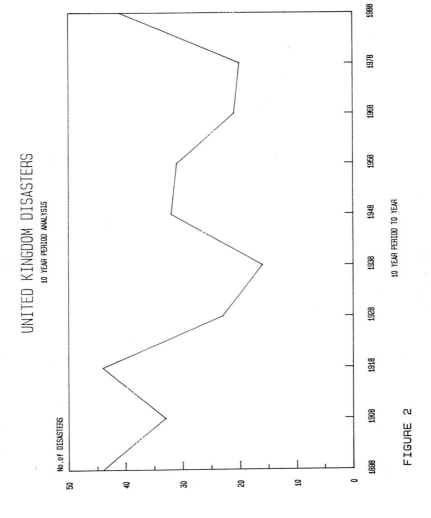

UNITED KINGDOM DISASTERS

10 YEAR PERIOD ANALYSIS

No. of DISASTERS

10 YEAR PERIOD TO YEAR

FIGURE 2

9

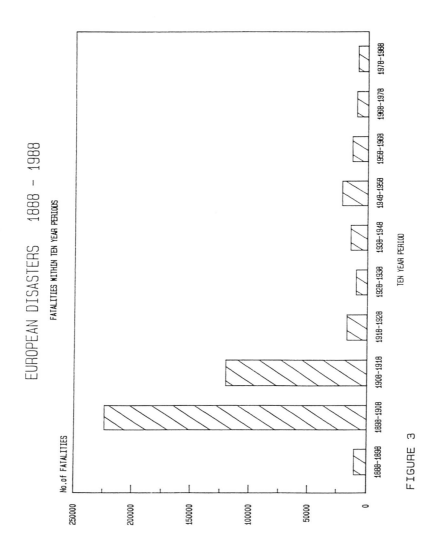

EUROPEAN DISASTERS 1888 - 1988

FATALITIES WITHIN TEN YEAR PERIODS

No.of FATALITIES

250000

200000

150000

100000

50000

0

1888-1898 1898-1908 1908-1918 1918-1928 1928-1938 1938-1948 1948-1958 1958-1968 1968-1978 1978-1988

TEN YEAR PERIOD

FIGURE 3

10

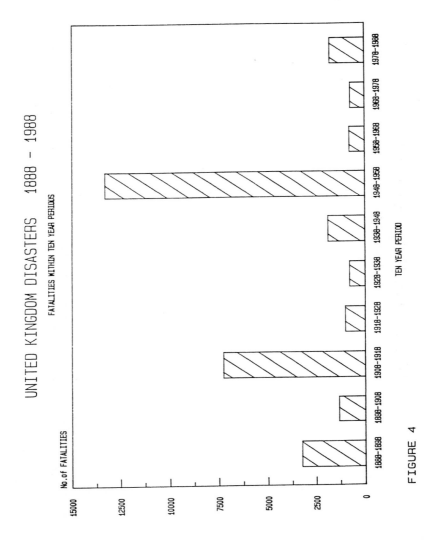

UNITED KINGDOM DISASTERS 1888 – 1988

FATALITIES WITHIN TEN YEAR PERIODS

FIGURE 4

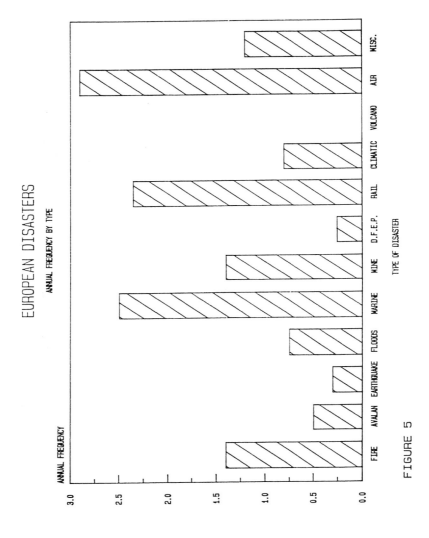

EUROPEAN DISASTERS

ANNUAL FREQUENCY BY TYPE

FIGURE 5

12

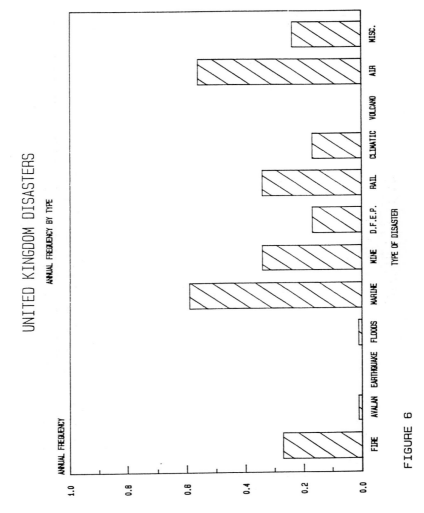

UNITED KINGDOM DISASTERS

ANNUAL FREQUENCY BY TYPE

FIGURE 6

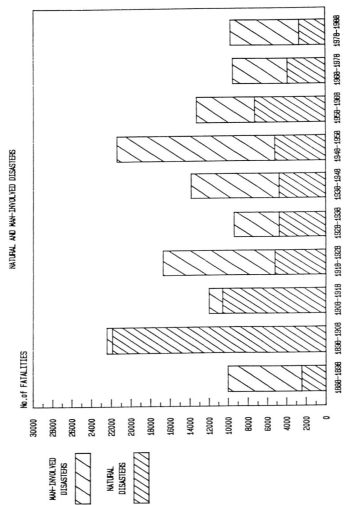

EUROPEAN DISASTER FATALITIES 1888 - 1988

NATURAL AND MAN-INVOLVED DISASTERS

FIGURE 7

14

categories).The results for the U.K. are the reverse of what has happened on the European mainland,man-made and hybrid deaths far exceeding those occurring in natural disasters,(Figure 8.)

This is due to the fact that the U.K. is relatively immune from many major natural disasters such as earthquakes,avalanches and landslides.

The Bradford Disaster Scale and Classification.

Due to emotional and other factors there is often difficulty in comparing one disaster with another.Magnitude of fatalities is obviously the most important factor,and to a lesser extent the type and origin of the cause.Because of the large variation in numbers of fatalities that can occur in a disaster,this range is typically 10 to 1,000,000,the human mind has often difficulty in the perception of the magnitude and scale of disasters.

For this reason the following Scale of Magnitude is proposed

TABLE 2

Bradford Disaster Scale

No.of Fatalities	Magnitude
10	1
100	2
1000	3
10000	4
100000	5
1 million	6

Intermediate values of magnitude are given simply by the common logarithm (base 10) of the number of fatalities.
This methodology is similar to that use by Richardson(3),and reference to it is made in Marshall(4).
Other studies of major incidents include Fernandes-Russell(5),Fryer and Griffiths(6),Griffiths and Fryer(7) and Grist(8).

Complimentary to Table 2 a classification scheme can be introduced such that,

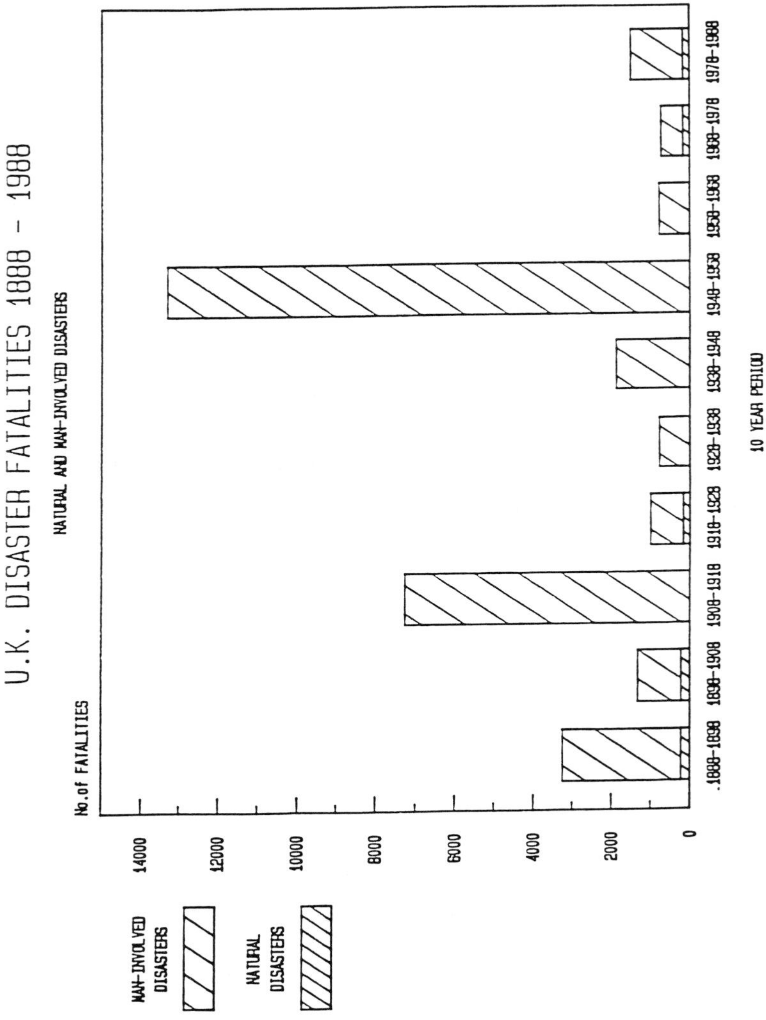

U.K. DISASTER FATALITIES 1888 - 1988
NATURAL AND MAN-INVOLVED DISASTERS
No.of FATALITIES
14000
12000
10000
8000
6000
4000
2000
0
1888-1898
1898-1908
1908-1918
1918-1928
1928-1938
1938-1948
1948-1958
1958-1968
1968-1978
1978-1988
10 YEAR PERIOD
MAN-INVOLVED DISASTERS
NATURAL DISASTERS

FIGURE 8

TABLE 4

Fatalities	Class
$10 - 10^2$	1
$10^2 - 10^3$	2
$10^3 - 10^4$	3
$10^4 - 10^5$	4
$10^5 - 10^6$	5

In Table 5 examples are presented using this scale for some disasters that have occurred recently ,

TABLE 5

Notable Disasters By BDS Magnitude and Classification

Disaster	Fatalities	Magnitude	Class
Herald of Free Enterprise	188	2.27	2
Piper Alpha	166	2.22	2
Armenian Earthquake	24,000	4.38	4
Hillsborough	95	1.98	1
Clapham Common	36	1.56	1
1988 Bangladeshi Floods	2 million (approx.)	6.3	6

Using the Bradford Disaster Scale and Classification one can now analyse the frequency of disasters that have occurred in the U.K. and on the European mainland over the last one hundred years.

TABLE 6

Frequency of Disasters
by BDS Class
1888 - 1988

BDS Class	Europe	U.K.
1	901	237
2	217	44
3	19	1
4	4	1
5	1	0

Hundred year frequencies of magnitudes of disaster that have occurred within Europe is presented in Figure 9,which has been derived from Table 6.The majority of disasters are of a magnitude less than 3 (Table 6),the pattern is similar for the U.K..As is shown in Figure 10.

The results of plotting the logarithm of the frequencies of occurrence of disaster in the last one hundred years for Europe and the U.K. against the class are shown in Figures 11 and 12 respectively.Both plots result in approximately straight lines of similar slope.

PROBABILITY OF OCCURRENCE.

For emergency planning purposes it would be useful to know not only the probability of a disaster occurring in a given time interval but also the likely magnitude of that disaster.

The following Figures give the probability of certain numbers of disasters occurring in a given year for class 1 and 2 disasters for Europe and the U.K.respectively.For class 1 European disasters,Figure 13,the probability of no disaster occurring within a particular year is extremely small.The most likely scenario is that within any year Europe will experience between 5 and 11 class 1 disasters.For disasters of class 2 the highest probability is that at least two incidents will occur each year which will claim between 100 and 1000 lives.(Figure 14)
In Figures 13 and 14 observed frequencies are compared with those predicted using Poisson distribution.

Within the U.K. for class 1 disasters there is a high

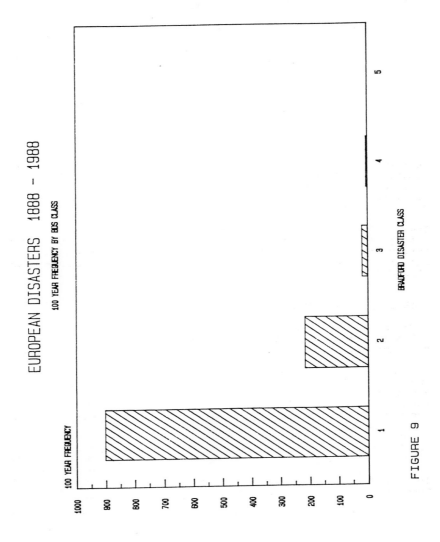

EUROPEAN DISASTERS 1888 – 1988

100 YEAR FREQUENCY BY BDS CLASS

100 YEAR FREQUENCY

BRADFORD DISASTER CLASS

FIGURE 9

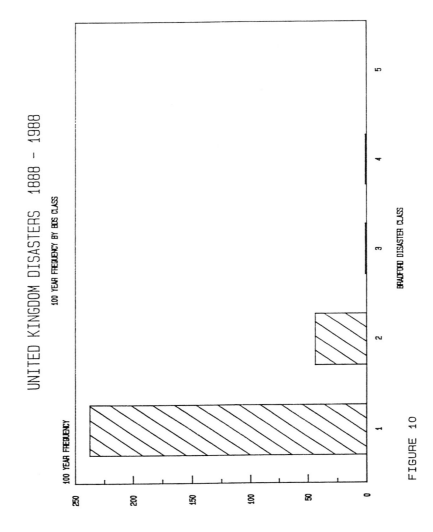

UNITED KINGDOM DISASTERS 1888 – 1988

100 YEAR FREQUENCY BY BDS CLASS

FIGURE 10

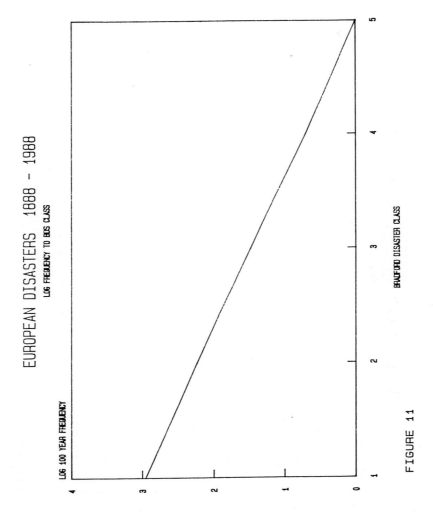

EUROPEAN DISASTERS 1888 - 1988

LOG FREQUENCY TO BDS CLASS

LOG 100 YEAR FREQUENCY

BRADFORD DISASTER CLASS

FIGURE 11

21

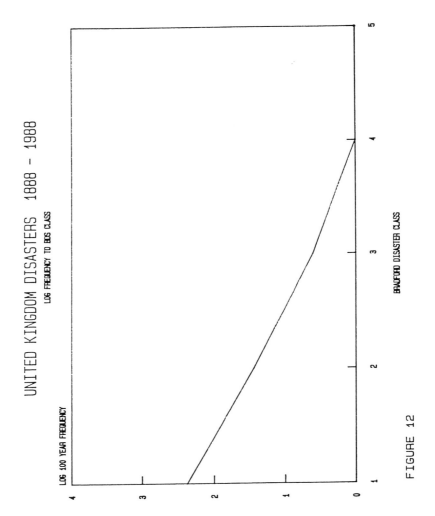

UNITED KINGDOM DISASTERS 1888 – 1988

LOG FREQUENCY TO BDS CLASS

LOG 100 YEAR FREQUENCY

BRADFORD DISASTER CLASS

FIGURE 12

EUROPEAN DISASTERS   BDS CLASS 1

PROBABILITY OF FREQUENCY PER ANNUM

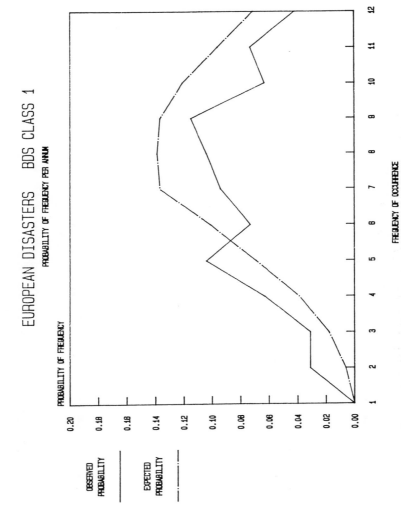

FIGURE 13

23

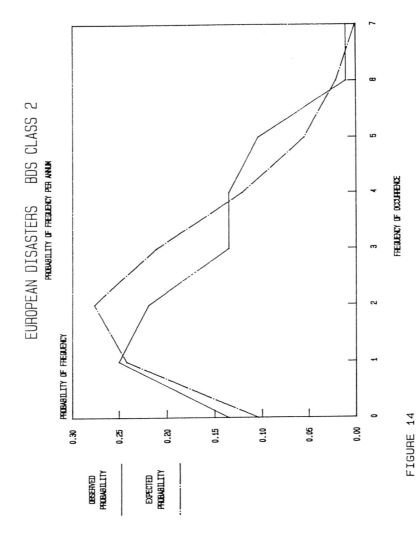

EUROPEAN DISASTERS   BDS CLASS 2
PROBABILITY OF FREQUENCY PER ANNUM

FIGURE 14

24

probability that there will be at least two class 1 disasters
each year and there is a reasonable probability that there
could be three or four , (Figure 15). For class 2 disasters the
probability is that one such incident will occur every three
to four years. (Figure 16)

TABLE 7

Summary of Disaster
Frequencies

BDS Class	Europe Frequency p.a.	U.K. Frequency p.a.
1	9	2
2	2	0.33
3	0.2	0.04
4	0.05	0.0067
5	0.01	0.001

Frequency Analysis by Type of Disaster

As discussed earlier, Europe and the U.K. differ in that the
U.K. has a low rate of natural disasters, which cause the
greatest loss of life, compared to Europe generally.

Within Europe the disasters with the highest probability of
occurrence are those of man-made origin, i.e. air, rail, marine
and fire. The cumulative frequency of these four types is
approximately 0.7 per year.

Two periods have been analysed, 1938 - 1963 and 1963 -
1988, for Europe and the U.K. to see if any change in the
pattern of frequencies of occurrence has taken place; results
are given in Figures 17 and 18 respectively.

The largest change is in those disasters that have occurred
within Europe (Figure 17) resulting from mining incidents
where the frequency has dropped by over 75 %. The frequency of
marine disasters has dropped by around 50% due to the shift
away from passenger sea travel to air travel. The small
increase in the frequency of air disasters when compared to
the considerable increase of usage of this means of travel
clearly demonstrates the very large advances in air safety
that have occurred over the last fifty years. The numbers of
fatalities per disaster for air disasters has however
dramatically increased due to the much larger aircraft now in

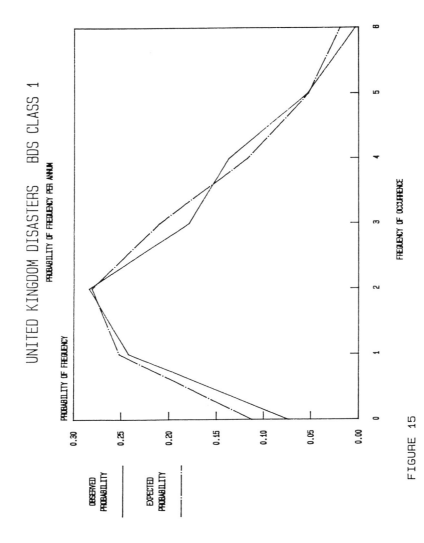

UNITED KINGDOM DISASTERS   BDS CLASS 1

PROBABILITY OF FREQUENCY PER ANNUM

FIGURE 15

26

UNITED KINGDOM DISASTERS   BDS CLASS 2

PROBABILITY OF FREQUENCY PER ANNUM

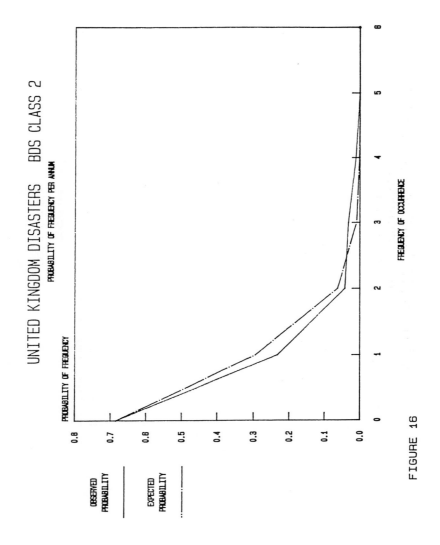

FREQUENCY OF OCCURRENCE

FIGURE 16

27

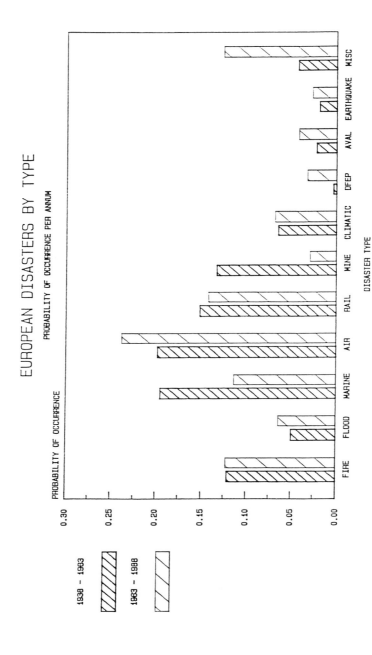

EUROPEAN DISASTERS BY TYPE

FIGURE 17

use and the greater seat capacity.However,this is more than balanced by the dramatic drop in the frequency of accidents per flight and better designed aircraft operating within much improved flight rules and regulations.

Natural disasters have maintained a reasonably constant frequency over the two periods of analysis.

A similar analysis has been carried out for U.K. disasters for the same periods,Figure 18.The frequency of air disasters has remained constant whereas marine and mine disaster frequencies have dropped dramatically.Marine travel has decreased very significantly as a form of transport over the past forty years.The decrease in mine related incidents has occurred from the greater use of safety technology and a much reduced manpower levels within the industry.The frequency of fire related disasters has increased by 100% and the climatic disasters have risen by a factor of four.Large increases in frequencies have occurred within the DFEP and Miscellaneous groupings.Over the past twenty five years there has been a doubling of the number of small scale infectious outbreaks,especially within residential homes for the elderly.The Miscellaneous group contains those elements due to road transport accidents and within the past ten years there has been an increase in the number of large group fatalities especially on the motorways.

Fatalities by Disaster Type Analysis

For the period 1963 - 1988 the average number of fatalities for different types of disaster which have occurred within the U.K.have been analysed and the results presented in Figure 19.Only one major landslide,Aberfan,has occurred during this period and which claimed a total of 144 lives.The others are more representative of their groups.

Marine and air disasters have produced the largest number of fatalities over the last twenty five years although the marine results are distorted due to the large number of lives lost with The Herald of Free Enterprise.The marine fatalities would have been in the region of 20 for the above period of time.Air disasters average around 75 and rail disasters around 35 fatalities per disaster.

Fatalities per Incident

In this section an analysis of average fatalities are studied as a function of disaster magnitude of disasters occurring within the U.K..

Class 1 European disasters,Figure 20,show that the highest probability is that between 10 - 30 lives will be lost per disaster.For Class 2 European disasters,Figure 21,6 out of every ten disasters will claim 100 - 200 lives and 9 out of every 10 will result in the loss of between 100 and 400 lives.

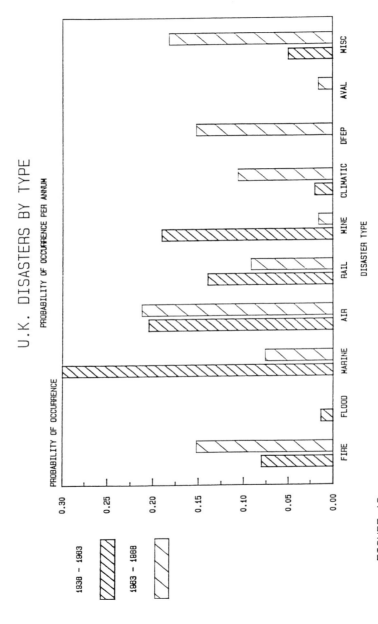

FIGURE 18

30

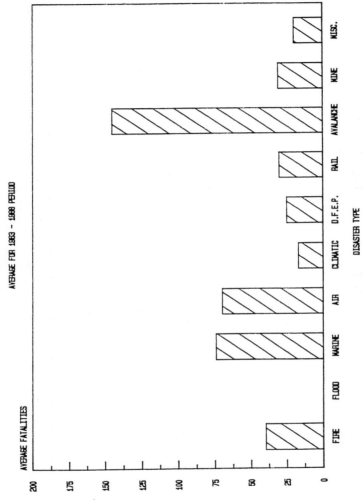

U.K.FATALITIES BY DISASTER TYPE

AVERAGE FOR 1963 - 1988 PERIOD

FIGURE 19

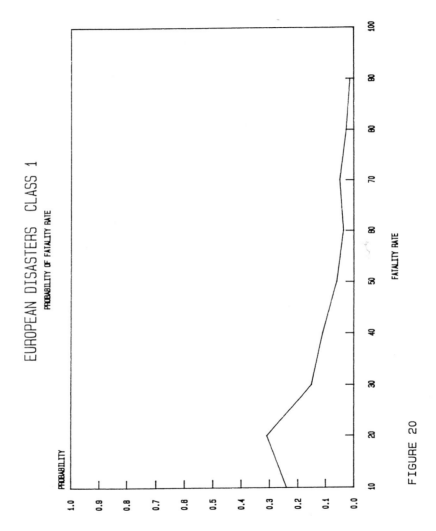

EUROPEAN DISASTERS CLASS 1
PROBABILITY OF FATALITY RATE

PROBABILITY

FATALITY RATE

FIGURE 20

32

For class 1 disasters occurring within the U.K.the probability of number of fatalities is highest between 20 - 30 fatalities per disaster and for up to forty fatalities per disaster the probability rises to 0.7.(Figure 22)

There is a probability of just below 0.8 that any European disaster will be of class 1 - 2,Figure 23,and a probability just greater than 0.2 that it will have a magnitude greater than 2.
For the United Kingdom, on the occurrence of a disaster,the probability of the number of fatalities being of magnitude 1 - 2 is about 0.83 and for a class 2 disaster the probability is around 0.16.(Figure 24) leaving a probability of 0.01 for disasters of class 3 or greater i.e. more than 1000 deaths.

Although emphasis has been placed on the number of fatalities that occur in a disaster,the number of injured involved can be an order of magnitude higher.

For planning purposes it is suggested that further research be carried out to establish the likely numbers of injured in a disaster of particular magnitude and type.

Statistics giving the number of injured survivors are not readily available and those that are often do not lend themselves to easy interpretation.
As a general rule for each fatality there are between 3 and 10 injured survivors with differing degrees of severity of injuries.A rail crash seldom kills all on board,the injured are mainly in the same carriages where the fatal injuries occurred.An air disaster occurring during mid flight,which is a relatively rare occurrence,produces no survivors whereas disasters occurring on take off or landing have about a 25% survivor rate with another 25% suffering severe injuries.

DISCUSSION.

The problem of classification of disasters to allow comparisons between different types has been studied and some key factors identified.These key factors are,number of fatalities,type of disaster and the degree of man-involvement.

Within Europe there is a high expectation that between 5 and 11 class 1 disasters will occur each year.Each of these class 1 disasters is likely to produce between 20 and 50 fatalities.Furthermore,there is a high likelihood that they will arise from man-made sources and will involve either air ,rail or marine transport or be the result of the effects of a major fire.
The frequency of class 2 disasters will be around 2 to 3 per annum with an equal probability that it will result from natural or man-involved sources.

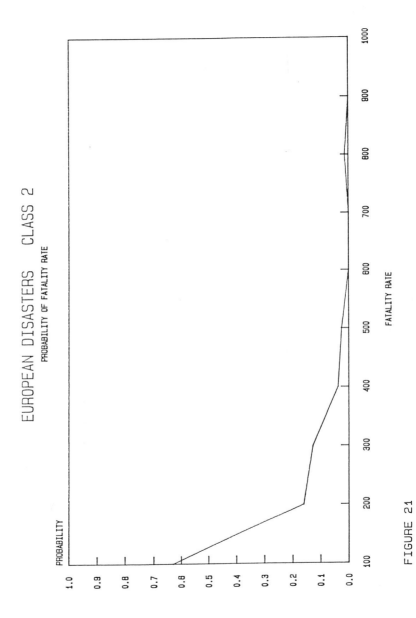

EUROPEAN DISASTERS   CLASS 2
PROBABILITY OF FATALITY RATE

FIGURE 21

34

UNITED KINGDOM DISASTERS CLASS 1

PROBABILITY OF FATALITY RATE

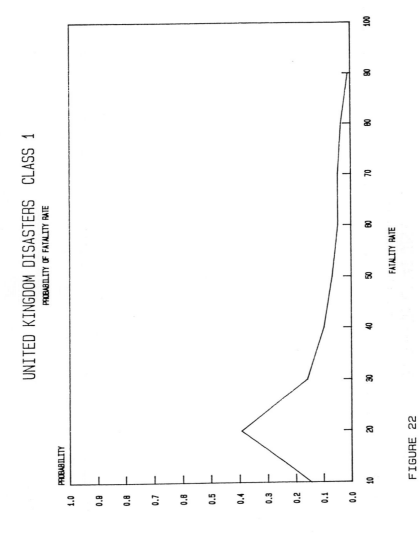

FIGURE 22

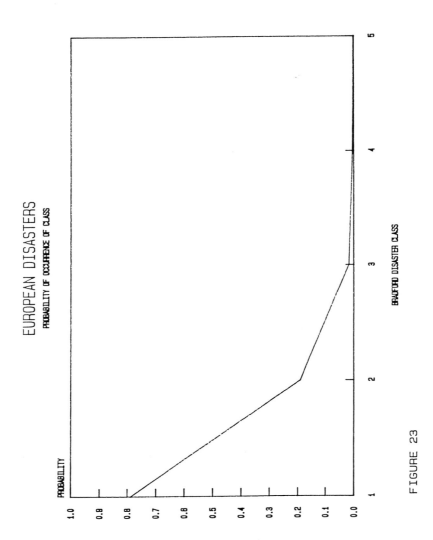

EUROPEAN DISASTERS

PROBABILITY OF OCCURRENCE OF CLASS

PROBABILITY

BRADFORD DISASTER CLASS

FIGURE 23

36

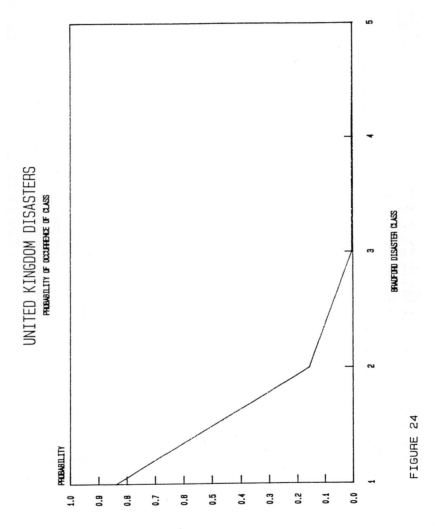

UNITED KINGDOM DISASTERS

PROBABILITY OF OCCURRENCE OF CLASS

PROBABILITY

BRADFORD DISASTER CLASS

FIGURE 24

37

Class 3 or above disasters can be expected to occur once in every 3 years and will be of a natural origin.

For the United Kingdom the scenario is somewhat different in that two class 1 incidents can be expected to occur every year and there is a significant probability that this number could increase to up to five in any one year.Class 2 disasters can be expected once every 3 years,although in the last two years in the U.K. there have been three class 2 disasters.Pure statistical analysis at present is unable to establish whether the higher number of class 2 disasters which have occurred within the last two years are random rare events or are the beginning of an upward trend arising from some yet unidentifiable structural societal change such as undermanning

As regards overall planning strategies the present study suggests that emergency planning could perhaps mainly concentrate on preparing for disasters of man-made origin and in which fatalities are between 20 and 40 with up to 100 survivors with severe injuries requiring immediate medical treatment.These estimates are of necessity of a very approximate nature.However,it is believed that the present study does provide a methodology and a classification system for analysing disasters so that planning on a regional or national basis regarding provision of necessary and relatively scarce resources can be made on a quantitative judgemental basis.

CONCLUSIONS

1).A quantifiable system of classification has been demonstrated.

2).There has been an increase in the U.K. in the number of disasters over the past ten years of over 30% .

3).There is a high probability that a U.K. disaster will be either of transportation or fire origin.

4).The likelihood of a major natural disaster occurring within the U.K. is small.

5).The majority of disasters that occur in the U.K. are of anthropogenic origin and in principle are preventable.

RECOMMENDATIONS.

1). That the usefulness of the method for emergency planning purposes be assessed.

2). That the method be extended to include incidents that could have resulted in potential disasters.

3). That the method be used for initial hazard identification and risk evaluation and as a tool for strategic planning.

4). That the scale when combined with the probability of occurrence be developed as a tool for resource allocation and level of response planning.

5). That the research be extended to include factors in the scale other than fatalities;i.e.injuries,economic and long term effects etc..

References

1). The Oxford English Dictionary,1966 (iii),Claredon Press.

2). Nash,J.R."Darkest Hours."1971,Prentice Hall.

3). Richardson,L.F.,(1960),"Statistics of Deadly Quarrels",Atlantic Books(London)

4). Marshall,V.C.(1987),"Major Chemical Hazards",Ellis Horwood

5). Fernandes-Russel,D.,(1987),Environmental Risk Assessment Unit Report No.2,University of East Anglia

6). Fryer,L.S.and Griffths,R.F.(1979),United Kingdom Atomic Energy Authority Report SRD R 149,UKAEA

7). Griffiths,R.F. and Fryer,L.S.(1978),United kingdom Atomic Energy Authority Report SRD R 110,UKAEA

8). Grist,D.R.,(1978),United Kingdom Atomic Energy Authority Report SRD R 125,UKAEA

EEC INVOLVEMENT IN DISASTER PREVENTION

P. D. Storey, Directorate-General for Science,
Research and Development Commission of the
European Communities, Brussels

I. INTRODUCTION

The past 10 to 15 years has seen a growing awareness of the
catastrophic accident potential of the chemical and petrochemical
sector. The term "Major Technological Hazards" describes these
complex array of situations in process plants, storage,
transportation and delivery systems. Much attention is given to
safety in operation by the industry, but with varying degrees of
seriousness incidents do occur. Some 300 to 400 such accidents
happen worldwide each year and chemical disasters rank fairly high
among these in terms of loss of life, financial loss or environmental disruption.

The early 1970's saw a major impetus given to research in this
subject, firstly in the USA, with concern for massive liquid
natural gas imports, and then in the UK and elsewhere following the
Flixborough disaster. At first priority was given to researching
fundamental chemical and physical phenomena such as dense gas
dispersion and combustion. Now, in addition to looking for ways of
preventing and mitigating the consequences, attention is being
focused on human and societal aspects. The planning, communication
and emergency response for accidents and the perception of risk and
its assessment have all become issues of major interest.

2. MAJOR TECHNOLOGICAL HAZARDS

The term major technological hazards describes certain dangerous
substances having the potential to give rise to serious injury or
damage beyond the vicinity of the workplace but which occur as a
result of industrial technology rather than naturally occurring
phenomena such as earthquakes. In the main such major hazards in
the chemical and petrochemical sectors are associated with four
hazardous inventories, which can be grouped as follows :

Note : The opinions expressed in this paper are those held by the
author.

1) Large quantities of flammable/explosive substances, mainly hydrocarbons and derivatives;

2) Large quantities of unstable or very reactive substances, peroxides, nitrates, liquid hydrogen and oxygen;

3) Large quantities of common toxic chemicals, chlorine and ammonia;

4) Small quantities of very toxic and persistent chemicals.

3. REGULATIONS

It was on these lines that the EC Directive "on the major accident hazards of certain industrial activities" (Seveso Directive) was drafted and adopted in 1982. In it are listed processes and substances with appropriate threshold quantities - an equivalent hazard concept. The directive deals only with stationary installations. Its formal requirements are :

1) Notification to appropriate authorities (above certain thresholds);

2) Preparedness, the implementation of internal and external emergency planning (above major thresholds).

In Great Britain the Control of Industrial Major Accident Hazards Regulations (CIMAH) implement the European Community Seveso Directive.

4. ON-GOING EC RESEARCH

The Commission implements a major accident reporting system (MARS) for use as a data base for information and for accident analysis. The identification and modification of these major hazards lead to the systematic safety assessment of potentially hazardous industrial sites - many methods have been developed but none universally accepted. Thus a primary motivation for EC research in this field lies in the existence and implementation of the Directive with the objective of obtaining a broad scientific-technical consensus in this complex field.

The overall aims of on-going EC-funded research addresses 3 primary issues :

1) Prevention, by appropriate design, and safety management;

2) Mitigation, by appropriate counter-measures and contingency/emergency planning;

3) Impact assessment, for siting of plants, storage facilities.

Research is carried out at the Joint Research Centre (JRC) and on a pilot scale through shared-cost contracts. The JRC concentrates on safety problems associated with runaway reactions and on risk assessment and risk perception in relation to be extended to human factors and emergency planning. Contract research deals with areas in which there is need for large-scale experimentation, field trials, wind tunnel simulations, extensive numerical modelling and involves transnational joint projects. The funding for this contract research was drastically cut from 15 to 3 MECU and some research areas had to be abandoned altogether. Out of 68 co-ordinated projects submitted only 5 could be retained.
The research addresses 3 primary topics :

1) Chemical and physical phenomena;

2) Risk assessment;

3) Risk perception.

5. NEW CONTRACT-RESEARCH PROGRAMME

However, over recent years there has been a vigorous renewed interest in major hazards research. Some serious accidents around the world involving the release of substantial quantities of toxic gases such as chlorine and ammonia and storage fires involving chemical stockpiles have again drawn attention to the related risks. Additionally, the disruptive power of confined explosions such as on Piper Alpha have all brought about the reorientation of research priorities. Moreover, increased attention is being paid to human factor involvement in accidents and there is also and increased interest in the overall management of risk and the planning of emergency procedures.

Following extensive consultation procedures the new programme has been written with three main themes in mind. Firstly, research into the chemical and physical phenomena of the accidental release of volatile liquids and gases which are toxic or flammable. The primary issue here is the assessment of risk to the public-at-large and the environment from such events as toxic gas clouds, confined and unconfined explosions and toxic products from fires. Secondly, research into technologies of accident prevention which would be aimed at producing measures to mitigate the consequences of an accident and to improve the intrinsic safety of plants and transport for carrying dangerous bulk substances. Thirdly, research into evaluation and management of risk which would be targeted at identifying gaps in knowledge, validating models and producing expert systems for risk assessment so as to establish and optimise the connections between plant reliability, human reliability and economic and social cost.

6. CONCLUSIONS

The EC has two distinct goals in disaster prevention. One goal is to put in place such regulations which will ensure that major hazards are identified. Having identified them, then to evaluate both the on-site and off-site risks and having evaluated them, then to notify the public close-by of the hazards and risks and to prepare emergency plans in case of disaster. The other goal is to generate fundamental research in all aspects of major technological hazards so that there may be a greater understanding of industrial hazards and the way they are managed not only in the interests of the European Community but to the world-at-large.

7. BIBLIOGRAPHY

Bourdeau, P., Green, G.ed. Methods for Assessing and Reducing Injury from Chemical Accidents, John Wiley 1989.

Council Directive 82/501/EEC Major accident hazards of certain industrial activities, 1982.

Sixth International Symposium, Loss Prevention and Safety Promotion in the Process Industries, Oslo 1989.

A guide to the Control of Industrial Major Accident Hazards Regulations 1984, HMSO 1985.

MAJOR ACCIDENTS

FEYZIN, FRANCE (1966)   Hydrocarbons

FLIXBOROUGH, UK (1974)   Hydrocarbons

BANTRY BAY, EIRE (1979)   Hydrocarbons

MEXICO CITY (1983)   Hydrocarbons

CERRITOS, MEXICO (1981)   Chlorine

LOS PAJARITOS, MEXICO (1984) Ammonia

BHOPAL, INDIA (1984)   Methyl Isocyanate

SHARED-COST RESEARCH TOPICS

BUDGET - 3 MILLION ECU -

1. Chemical and Physical Phenomena

1.2. Dispersion of gases and vapours

1.3. Flame Propagation in unconfined vapour clouds

1.4. Catastrophic Fires.

2. RISK ASSESSMENT

Benchmark exercice on major hazard analysis of an ammonia manufacturing plant.

3. RISK PERCEPTION

Observation of social response to major risks

NEW SHARED-COST RESEARCH PROGRAMME

BUDGET - 15 MILLION ECU -

Research in this area addresses 3 primary issues :

- Accident prevention, by appropriate design and safety management of processes and transport standards,

- Mitigation by appropriate counter-measures and contingency/emergency planning,

- Impact assessment for siting of plants, storage facilities, etc., safety auditing and monitoring.

Research complements work of Joint Research Centre.

In some areas research to be a continuation of pilot programme.

CHEMICAL AND PHYSICAL PHENOMENA

Assessment of risk to the public-at-large and the environment.

SOURCE TERM

Produce data on rate of release and models describing emission.

DISPERSION

Produce data on manner in which gases spread and models describing
dispersion.

COMBUSTION AND RELATED EFFECTS

Develop models to predict the effects of large flames or fires on the environment.

Investigate the damage potential of blast waves.

TECHNOLOGIES OF ACCIDENT PREVENTION

Produce measures to mitigate the consequences of an accident and improve the intrinsic safety of plant and transport.

EVALUTATION AND MANAGEMENT OF RISK

Research aims at identifying gaps in knowledge, validating models and producing expert systems for risk assessment.

1)   Hazard analysis

2)   Management of risk

3)   Human factors.

HAZARD PREDICTION

J. J. Clifton, Head of Major Hazards Unit,
AEA Technology, Safety and Reliability Directorate

SRD was founded in 1959 as the UKAEA's own watch-dog on safety at its experimental establishments. It was in the nuclear context that SRD first began to develop risk assessment techniques in the 1960's and early 70's. As early as 1965 SRD began applying techniques for systematic analysis of possible failures to other industrial activities, and in the early 70's began work on the world's first complete risk assessment of a chemical/petrochemical complex. This, the Canvey Island Study, was published in 1977.

Since that time, SRD's Major Hazard unit has been the focal point for a substantial research programme funded by the Health and Safety Executive, and for an increasing range of projects carried out for industry on a consultancy basis. As demands for safety from the public and regulatory bodies grow, Hazard prediction is becoming more and more a part of the management of potentially hazardous plant and activities - to demonstrate compliance with standards, to help optimise investment for safety and productivity, and to provide information for emergency planning purposes.

The paper defines risk and risk assessment and identifies the four major components to the methodologies used in the assessment of risk. These four components are discussed and the various techniques that have been evolved are described. Current developments are identified and possible topics for future research discussed. It concludes by considering the applications and benefits of probabilistic risk assessment.

Before going on to discuss the topic of risk assessment it is worth spending a few moments to discuss the nature of the problem which makes risk assessment so necessary.

Whilst the chemical industry has been growing steadily from the start of this century in the last 30 years or so a number of factors have combined to exacerbate and highlight the problem of major chemical hazards. Firstly the improved understanding of scale up of chemical processes together with better materials and fabrication technology meant that during the 60's and 70's unit sizes of chemical plants increased significantly - in some cases up to 20 times - with the consequential economies of scale but, of course, with inventories of products and intermediates increased correspondingly together with the consequence of any potential major accident.

Secondly, new processes were developed and the ranges of chemicals, especially those derived from petroleum products, widened significantly. The technology to liquefy and store petroleum and other gases such as methane, and ethylene in large quantities was established. A third factor has been the increasing concern over the period both nationally and internationally about the environment and public safety leading to increasingly stringent requirements on the process industries. In the UK and probably elsewhere this increased public concern has drawn attention to the evolving regulatory situation and caused an upsurge in the interest paid to public inquiries, planning consents and emergency arrangements.

Thus, whilst the likelihood of an accident at a particular plant has been reducing with more plants and greater inventories of toxic or hazardous materials, the consequences of any individual major accident have been steadily increasing. It is not necessary to dwell for long on the need for "safe" chemical plants. Apart from the potential for harm to the public an accident can have major economic implications for the operator with the loss of the plant and the need for compensation as well as having an adverse effect on the public image of the company. All companies therefore have a major incentive to prevent accidents; it is worth spending considerable sums of money to achieve this aim. A major question to answer is "GIVEN OUR COMMITMENT TO PREVENTING ACCIDENTS WHERE DO WE BEST PUT OUR MONEY TO ENSURE THIS OBJECTIVE"?

An attempt to answer this question is one of the main aims of risk assessment. The assessment of risks is something which we all do from time to time. For the most part these assessments are performed almost subconsciously as, for example, when having looked to the left and right we decide that it is safe to cross the road. This does not mean there is no traffic on the road but rather that we have determined on the basis of experience that the risk of being struck by one or more of the vehicles we can see in the distance is, to us, acceptably low. Insurance companies are also heavily involved in risk and determine the relationship between an insured sum of money and their insurance premium from statistics concerned with historical evidence and forecast trends.

The traditional approach to safety in the process industries has been to design on the basis of experience, to devise operating precautions and to develop codes of practice for applying experience to future designs. Recently many organisations have concluded that, although an essential ingredient, this retrospective view alone is not enough. These organisations now believe that it is necessary to try to anticipate what hazardous situations could arise and, having done this, that it is useful to express the chance and severity of such accidents on a quantitative scale in order to take decisions about what course of action to take. Techniques pioneered and developed in the nuclear and aerospace industries are now finding applications in the process industries although a number of organisations remain very sceptical about the usefulness of the quantitative approach.

Before defining the term Risk Assessment it is perhaps useful to define the word Risk. It used to be quite common for the words Risk and Hazard to be used as though they were interchangeable. There is now a growing consensus in the United Kingdom as to the general sense in which these terms are to be used. In the early 80's a working party, made up from practising engineers from industry, consultants, universities and the authorities, was set up to establish a set of terminology in the field of Hazard and Risk Assessment. This working party reported to the Engineering Practice Committee of the Institution of Chemical Engineers and in 1985 produced the booklet 'Nomenclature for Hazard and Risk Assessment in the Process Industries'. In this booklet Hazard is defined as 'a physical situation with a potential for human injury, damage to property, damage to the environment or some combination of these'. Risk is defined as 'the likelihood of undesired events and the likelihood of harm or damage being caused together with the value judgements made concerning the significance of the results'.

Risk assessment is aimed at answering two questions:

How bad could it be?

How often might it happen?

This is regarded as the public perspective, industry is more likely to ask:

How can I minimise the effects? or

How can I reduce the chance of it happening?

The same techniques of risk assessment can be used to develop answers to all these questions and it is the techniques and their use that this paper is all about.

This review is not exhaustive - the aim is to draw attention to some of the more common methods in use and to promote discussion on their strengths, weaknesses and position in overall risk management.

Risk assessment methods can be broken down into four main areas:

methods used to identify sources of accidents and the ways in which they could occur

quantitative estimation of the likelihood of occurrence of these accidents - known as the frequency analysis stage. Here one would try to estimate whether the undesired event was likely to occur every ten years or every million years or whatever

quantitative estimation of the potential consequences of accidents - known as the consequence analysis stage. Here we would try to estimate the probability that people located in different environments at different distances from the scene of the undesired event would be killed or seriously injured the calculation of risk levels - these are calculated by combining the frequency and consequence estimates. The risks would normally be expressed in terms of the likelihood of death or serious injury to members of the workforce and adjacent population but could also be expressed in terms of parameters such as resulting financial loss. This is also a sort of 'so what' stage in which an assessment is made of the acceptability of the risk levels estimated together with the possible need for remedial action.

Risk assessment differs fundamentally from the 'deterministic' approach in that there are no preconceptions as to the credibility of any type of accident. Any hazard or accident scenario that can be identified is included for analysis. For example, this would include accident scenarios arising from human error, equipment failure and external hazards such as aircraft crashes. Scenarios are only subsequent discounted if it can be shown that either:

1. The consequence of its occurrence in terms of their effect on adjacent population is negligible or

2. The likelihood of their occurrence is negligible when compared to the likelihood of occurrence of other accident scenarios that have been identified.

The particular strength of the risk assessment technique is that being quantitative in nature it provides a more explicit statement of the risks associated with a particular activity than qualitative generalizations. As such this should help to avoid sterile arguments where the proponents of a potentially hazardous activity content themselves with statements that they consider it safe, that it conforms with Codes of Practice and hence the likelihood of a serious incident can be effectively discounted whilst their opponents voice diametrically opposite views. The very nature of risk assessment forces both sides to try to develop their arguments to a much greater depth with a view to being able to put forward detailed quantitative analyses which will withstand close scrutiny. At the end of the day whilst there still may not be agreement about the absolute magnitude of some of the risks, there should be a much better understanding of the system and its potential weaknesses and this can then lead to the identification of possible modifications to strengthen the weak links and hence significantly reduce the overall risks.

The first area of risk assessment is hazard identification which uses three categories of techniques: those that adopt a comparative methodology, those that apply a fundamental methodology and those that use failure logic diagrams.

The comparative methodology of hazard identification uses check lists and hazard indices. Check lists are essentially a simple and empirical means of applying experience to designs or situations to ensure that the features appearing in the list are not overlooked. A variety of check lists have been developed for application in the chemical and process industries covering many aspects of design, commissioning, operation and decommissioning. Lists may relate to material properties, they may be equipment specific (eg pipework fabrication) or they may be far more general plant with wide application such as auditing for hazards identification. Where major hazard accidents could occur, reliance on check lists alone would generally not be considered to produce a broad ranging or deep enough consideration of what could go wrong and how. Nevertheless they can have a useful role at various stage in a design to ensure that experience is taken into account and that basic general considerations are not overlooked. A check list will also serve as a list of subject pointers which will require attention at each stage in the life of a process. They are most effective when used to stimulate thought and enquiry through open ended questions rather than in the form that requires yes/no answers. For instance, having identified that overpressure may exist, by asking "How is the system protected against overpressure?" rather than "Is the system protected against overpressure?".

Hazard indices, like check lists, are empirical in nature. At least two versions have been published - the Dow Index and the Mond Fire, Explosion and Toxicity Index. Both indices take the same approach in that they are designed to give a quantitative indication of the relative potential for hazardous incidents associated with a given process unit.

Fundamental Methods are structured ways of stimulating a group of people to apply foresight in conjunction with their knowledge to the task of identifying hazards mainly by raising a series of "What if?" type questions. These methods have the advantage that they can be used whether or not codes of practice are available. There are two main techniques available in this family of methods, Hazard and Operability (HAZOP) studies and Failure Modes and Effect Analysis (FMEA). Both are structured to improve the completeness of hazard identification by applying knowledge through anticipation of dangerous situations. HAZOP is a technique for systematically considering deviations from the design intent by the application of a series of guide words such as too much, too little, etc, to the process parameters in search of undesirable process deviations.

FMEA is probably more used for reliability assessment than risk assessment but it can be used to identify initiating faults that could lead on to worse things. It normally works at a detailed plant level. Individual components or functional systems are considered and a list made of all possible failure modes.

Fault logic diagrams offer a pictorial method for the presentation of fault logic. They encourage the analyst to speculate how a particular situation could arise or what may ensue from a situation and hence identify causes or outcomes of undesired events. Their main purpose is to provide a structure to the failure logic and show the interdependences between events and system states. The main techniques adopting this approach are Fault Trees and Event Trees. These techniques both use binary logic, where the diagram lines connecting faults are on or off only.

Fault trees are best applied when seeking to define all possible routes to a failure of some sort. Event trees are best applied when looking to define how a failure might lead to grave consequences. One therefore normally seeks suitable nodal events that can comprise the result of a fault tree which is the beginning of, or a branch on, an event tree. In fault trees we think backwards - the starting point is the nodal event which is called the top event in fault tree terminology. The first step is to find a first level of faults that singly or in combination could lead to the top event. The next step is to look for all faults that could give rise to the first level faults and then so on down the tree until a point is reached that is some failure for which data is available. This is called a basic event. What can be learned from a fault-tree? There are both qualitative and quantitative aspects. One particular merit of the technique is that it reveals the way in which faults inter-relate. This qualitative use may be sufficient to indicate that no previously unforeseen combination of faults exist that will lead unexpectedly to the top event. Quantitatively, it opens the way to the second stage of risk assessment - the methods used to estimate the likelihood of occurrence of accidents.

There are two basic approaches to estimating the frequency of an accident. These are:

1. The use of historical data for similar accidents.

2. The synthesis of the accident frequency from a study of the combinations or sequences of events and failures which could cause the accident, together with data for their likelihood of occurrence.

Most risk assessments use a combination of these two approaches. Wherever possible failure data from the plant concerned, or from similar plant operating under similar conditions should be used. For comparatively rare failures this will not be possible and it is necessary to use generic data collected from a large sample. In this case major differences between the plant under study and the equipment from which the data has been gathered must be considered carefully. For some types of event the historical frequency figure is likely to be easier to estimate and more reliable and there is little point in trying to synthesize a figure. Even where figures are synthesized comparisons with historical experience should be made and reasons identified for any major differences in the results from the two methods.

The use of historical data has been greatly assisted by the creation of large databases which enable details of relevant past accidents to be analysed. One such database is MHIDAS (Major Hazards Incident Database Service) which was created by SRD in collaboration with the HSE. This database has over 4,000 records of incidents involving hazardous materials that resulted in, or had the potential to produce, a significant impact on the public at large.

The synthesis approach is generally based on the techniques of fault and event tree analysis which were described earlier.

There are three distinct stages in the third component of risk assessment which is that of estimating the potential consequences of accidents. The first is to determine the release mode, size and rate of the hazardous material concerned. The second is to determine the behaviour of the material after its release and the third is to consider the effects of the material on people.

To determine the mode, size and rate of the release it is first necessary to identify the vessel or pipework which normally contains the material. The next stage is to determine if the vessel or pipework is under pressure and, for the hazard under consideration, if it will fail catastrophically or just leak.

In the process industries, many gases are stored or handled as liquids under pressure or refrigeration or a combination of the two. The released material may therefore be as a non-flashing liquid, a flashing gas/liquid mixture or a gas. If a vessel containing liquefied gases under pressure fails catastrophically, the vessel contents will be released violently to the atmosphere. Gas will be formed and the material will cool to its boiling point. To determine the rate and amount released, the proportions of the contents initially released either as a gas or liquid can then be calculated.

If a non-pressurised vessel containing either gases liquefied under refrigeration or materials which are liquid at normal temperatures fails, the material will flow out to form a pool. For refrigerated materials the heat input from the surrounds will boil off some of the refrigerated liquid in a gas.

One point to remember is that many storage tanks containing non-pressurised liquids are bunded, that is surrounded by a low earth and/or concrete wall which should contain the tanks contents providing they are released slowly. However, for catastrophic failures it is known that as much as half the tanks contents may overtop the bund due to dynamic effects unless the bund has been specially designed to prevent this.

If the material is flammable it may be ignited at the point of release and either cause an explosion or burn as a torch or pool fire. Apart from the very large explosions, the main source of hazard to the surrounding populace will come from the dispersion of materials which are either gaseous or in fine respirable particulate form at normal temperatures and pressures and are either toxic or if flammable do not ignite instantaneously.

A major factor in the assessment of the hazards from dispersed gases is the density of gas cloud. If it is lighter than air it will rise and be dispersed by atmospheric turbulence. A gas heavier than air will linger near the ground, spread in a flat pancake shape. Under these conditions the atmosphere's dispersive powers are greatly reduced.

Many of the materials present on those industrial sites thought to pose a high risk, form denser than air mixtures, either because of their high molecular weight or because of their low temperature shortly after release. A number of models to predict the behaviour of dense gases following release to the atmosphere have been developed. The program DENZ has been developed by SRD on behalf of the United Kingdom Health and Safety Executive. DENZ is used to calculate the dispersion of gases from instantaneous releases such as after a catastrophic failure. A similar program CRUNCH has been developed for so-called continuous releases similar to the circumstances which exist following a pipe break. For a toxic release, DENZ and CRUNCH will integrate the exposure time and concentration at a variety of points to enable the hazard area to be mapped out. For a flammable release DENZ and CRUNCH will map out the limits of dispersion to a given concentration. A particular point to note is that the initial slumping phase following a dense gas release results in a substantial hazard range in all directions from the point of release.

Another program, GASP, has been developed to model the spread of a pool of a refrigerated liquid. For liquids with boiling points less than ambient temperatures, gas is released as the liquid takes heat from the surroundings and boils. GASP considers the heat input from the ground (or sea), the wind and the sun. For low boiling point materials such as LNG and propane, the heat input from the ground dominates until it cools down. GASP can predict the behaviour of instantaneous or continuous releases onto open or restricted surfaces.

Knowledge of the release mechanism and the subsequent fate of the released material, ie either as a large heat source, and explosion or atmospheric dispersal, enables calculations to be made of the effects of the materials at any distance from the release point at any time. Often the main interest lies in predicting distances from a particular occurrence at which the effect on people will be to cause death or serious injury. This may arise from a variety of effects which include the intake of toxic material exposure to high thermal radiation fluxes and damage from overpressures generated in an explosion. In the case of toxic materials the main concern from the emergency planning point of view is generally with the health effects of short term exposure to relatively high concentrations of important industrial materials rather than to chronic or occupational toxicology. The evaluation of the information on the physiological response effects observed following exposure times. Also one must not neglect those materials, which in themselves are safely non-toxic, are prone to give rise to toxic products during combustion. A typical example would be the production of toxic materials when polyurethane foams are burnt.

The effects on humans of thermal radiation are much better understood. These effects are a function of heat flux and exposure time. Studies indicate that people outdoors start to be at risk at thermal fluxes greater than 4kWm^{-2}. The heat fluxes necessary to cause third degree burns to bare skin for an exposure time of five seconds are about 70kWm^{-2}. Heat fluxes of this magnitude would ignite most forms of everyday clothing.

Explosion damage will result in exposure to thermal radiation and to pressure waves generated in the blast. People are fairly resilient to overpressures generated in an explosion. A peak overpressure of about 5 psi (34 kPa) may cause eardrum damage to some people whereas some 40 psi (276 kPa) is necessary to approach the threshold of mortality. Structures are much more vulnerable and most of the damage to people following an explosion comes from the effect of the overpressure on a structure they are in or near. At pressures between 1 and 3 psi (7 and 21 kPa), windows will break and walls collapse and at overpressures greater than 7 psi (48 kPa) there will be extensive structural damage.

When trying to determine the hazard range from the release of a flammable material, it is necessary to consider a number of different fire types such as pool fires and fireballs. Spills of oil and oil products can create a pool fire as can spills of refrigerated LNG or LPG contained in some way if ignited. Simple theory predicts hazardous heat fluxes at some distance from a pool fire whereas observation of fire fighters show them at work close in to the fire. Methods have been developed which take account of smoke obscuration and atmospheric transmissivity. Hazard ranges predicted by these methods are now much more in line with those actually found.

A rapid release of flammable gas which is ignited before mixing with air can cause a fireball. One method of producing a fireball is by a BLEVE. A BLEVE or Boiling Liquid Expanding Vapour Explosion can occur for example when a vessel containing liquefied gas under pressure is exposed to fire and fails due to a combination of increased internal pressure and weakening of the vessel material. The liquefied gas is released violently to atmosphere and, if flammable, is immediately ignited to give a fireball.

In addition to a fireball the effects of a BLEVE also include blast damage from the energy released when the vessel suddenly fails and from the missiles generated by rocketing fragments of this vessel. The generation of missiles reminds one of the fact that it is necessary to evaluate the possibility of interactions with adjacent installations arising from such failures. A typical example being damage caused by rocketing fragments from bursting pressure vessels.

It should be clear by now that a large number of factors have to be taken into account when assessing the consequences of a major accident. Fortunately we now have most of the techniques and understanding to enable us to predict hazard ranges with some confidence. However, as stated earlier, our understanding of the way major hazards are realised is by no means complete. The science of consequence analysis is a developing field of technology and research is continuing at a steady pace. In the UK this includes such research as the experiments on the dispersion behaviour dense gases which were carried out at Thorney Island by the National Maritime Institute on behalf of the Health and Safety Executive. Computerised modelling techniques are being developed and refined to study such topics as the behaviour of tanks containing flammable materials when engulfed in flames.

Many of the accidents involving materials commonly used in the chemical and petroleum industries which have caused serious loss of life have been those which have occurred when the material concerned has been in transit. Perhaps the best-known and most horrific was that which occurred at Los Alfaques in Spain in 1978 when a road tanker carrying 23 tons of propylene liquefied under pressure BLEVE'd near a camp site. The radiation from the fireball was extremely intense and over 200 people died as a result of being badly burnt. Most of the techniques developed to assess the consequences of releases of hazardous materials from chemical and petroleum process plants are also capable of predicting hazard ranges from materials in transit and can be used to assist in formulating general advice such as evacuation distances when preparing emergency plans.

The fourth component of risk assessment is concerned with combining probabilities with consequences in order to determine and present the results of the risk analysis.

A detailed risk assessment will cover a number of accident scenarios. For each of these scenarios there will be an estimated frequency of occurrence, but there may not be a single resulting set of consequences. For example, weather conditions and wind direction may have very important effect on the outcome of a large release of toxic gas. The number of people on or near the plant will largely depend on the time of day. These and many other factors can materially affect the consequences of accidents. A probability must be assigned to each influencing parameter which reflect its involvement in a particular set of accident conditions Then, an overall estimate of the likelihood of a particular accident and its consequences can be made.

A format has to be chosen for presenting the results of the analysis and this should be done at the outset of the study. Although other methods are in use, a very effective presentation is achieved by considering the following:

> 1. The "individual risk" ie the frequency at which a particular individual may be expected to sustain a given level of harm from the realisation of specified hazards.
>
> 2. The "societal risk" ie the relationship between frequency and number of people suffering from a specified level of harm in a given population from the realisation of specified hazards.

Whatever method is chosen, it is important that the presentation should be concise and intelligible while also being open to scrutiny and providing clear information on what factors appear to be important in the overall analysis. This last point is important in providing evidence that the specific precautions taken to prevent accidents are appropriate in a particular case, in order to support a claim that the duties to reduce risks as far as reasonably practicable have been discharged.

The methodologies of risk assessment have matured a great deal in the last few years, particularly where quantitative techniques of analysis are concerned. Nevertheless there is still a great deal of work to be done in the further development or refining of existing methodologies. In addition there are many new areas of work which need to be tackled. Current developments and possible topics for future research are as follows:

(a) The development of models for determining source terms for the dispersion of hazardous materials.

(b) Creating databases to store and analyse data for topics such as the effects of incidents on the environment.

(c) Developing expert system based computerised models to deal with the large amounts of data, expert knowledge and judgement that form an inherent part of risk assessment.

(d) Understanding the sources and effects of those uncertainties which are an inherent part of probabilistic risk assessment.

(e) Trying to understand the interaction between human beings and complex systems.

(f) Developing methods of estimating and improving the reliability of software, particularly that used in the programmable electronic control systems found in modern plant.

The areas where some form of risk assessment has found application are steadily increasing. Some of the important areas are as follows:

1. Plant Design

Risk analysis can be used as an integral part of the design process starting from the earliest stages in the development of a project. There may be a number of different options and evaluation of these can be made by assessing whether some of the alternatives could lead to problems with risk levels. Also the suitability of potential plant locations can be screened by carrying out some consequence analysis and then assessing the results in conjunction with some pre-determined level of acceptance.

2. Plant Safety Reviews

Many process plants handling potentially hazardous materials were built a number of years ago when the same level of attention was not necessarily paid to the possible size and frequency of interactions with adjacent population and other plants as is the case today. There are a number of pressures which require companies to establish if these plants require some form of modification or up-rating in order to conform with current safety standards or management policies.

Such studies will usually require an estimation of the risk levels presented by these plants as a precursor to determine which, if any, improvements are needed. It is often found that safety reviews on older plants do indicate that risk levels need to be reduced but that the measures required to bring about a worthwhile reduction are not as costly as the operators initially thought.

3. Regulatory Submissions

In a number of countries, for example those members of the European Community, it is now necessary to prepare a 'safety case' to demonstrate that the major hazard potential of the activities have been identified and the appropriate controls provided. The scope and content of the 'safety case' imply a format requiring the use of some form of risk analysis.

4. Review of Operating Restrictions

Outside of the immediate environment of the process industries there are a number of other industries and activities where a proportion of the operational procedures are determined on 'safety' grounds. Often these result in severe restrictions being placed on certain activities and this can produce significant economic penalties. However when the restrictions are examined their basis is not always clear and in many cases they seem to have developed in an ad-hoc manner in response to individual incidents. Risk assessment can plan an important role in trying to rationalise the need for these restrictions.

5. Siting and Planning Approval

It is now quite common for some authorities to require a form of risk assessment before granting planning permission for installations handling hazardous substances. Typical planning problems for which answers have to be found include:

(a) Should a new and hazardous process plant be permitted on an existing industrial site?

(b) Should a new housing estate or hospital/school etc be permitted near to an existing hazardous process plant?

(c) Should anything be done about an existing hazard near to existing communities?

Each of these questions essentially involves plant-to-community interactions. Since it can often be the case that the largest accident possible is so great that no practicable separation policy could completely eliminate the hazard to the community, it follows that some degree of risk has to be accommodated and the problem is to define and control it, and it is here that risk assessment has a role to play.

6. Emergency Planning

Emergency planning measures, such as provision for warning populations at risk, evacuation, etc, necessarily involve consideration of the areas likely to be affected, the extent of damage, the time available for countermeasures to be deployed and so on. This will clearly require the application of some form of consequence analysis. However, where there is a conflict of priorities for resources, as is usually the case, it may be appropriate for the plan to take heed of the probability of occurrence in addition to the consequences of the various scenarios in order to arrive at a balanced view. In such cases some form of frequency analysis would also be required.

7. Insurance and Liability Planning

The techniques of risk assessment may be applied to a range of financial calculations. In such cases scenarios must be constructed which may result in financial loss and then, using similar approaches to those employed in assessing the risk to people, the probability of such a loss occurring and its severity may be determined. In many cases there is a great degree of overlap between risk assessments to people and financial assessments since the events causing damage to property, shutdowns in operations etc, are often those same events that need to be considered, and so on. This will clearly require the application of some form of consequence analysis. However, where there is a conflict of priorities for resources, as it is usually the case, it may be appropriate for the plan to take head of the probability of occurrence in addition to the consequences of the various scenarios in order to arrive at a balanced view. In such cases some form of frequency analysis would also be required.

In conclusion it must be acknowledged that risk assessment is not a cheap process - experienced professionally staff must be involved. Since many of the benefits are associated with the knowledge obtained by carrying out a study, this implies, where justifiable, that companies should set up an area of expertise within their own organisations. The techniques have considerable potential, providing a logical approach and contributing to the basis for decisions. It is particularly important that those who consider the results of a risk assessment recognise that the strength of such an assessment lies not in the absolute value of the results obtained but in the relativity of other assessments carried out in a similar manner. A particular benefit of risk assessment is that it highlights the main contributor(s) to risk and identifies possible routes to reduction. Sensitivity analyses can be easily performed.

This paper represents the views of the author alone and does not necessarily represent those of the Safety and Reliability Directorate or any other organisation. However, I am indebted to my colleagues at SRD for assistance in providing material for this paper.

THE HOME SECRETARY'S ANNOUNCEMENT ON CIVIL EMERGENCIES AND THE FUTURE ROLE OF THE CIVIL EMERGENCIES ADVISER

R. Kornicki, Civil Emergencies Secretariat,
The Home Office

Introduction

Ladies and gentlemen, when I joined the Emergency Planning Division of the Home Office nearly two years ago the first thing that happened to me was to be given a train ticket to Bradford to come up for a Disaster Conference here.

A one-off event I understood and only a small part of my responsibilities in the Division.

A great deal has changed since then obviously, a great deal of work has gone on and my responsibilities have changed with it to the point where I and my staff are engaged wholly on the civil emergencies function.

A brief word about definitions if I can take up Dr.Keller's points here.

Within the content of a review that the Home Office has carried out, we define disasters not by the numbers of dead and, perhaps, it is a rash man who attempts any definition at all, but in relation to their demands on the local services and in essence here we broadly go along with the definition used by ACPO which defines a disaster as;

"an event which goes beyond the immediately available local resources and requires a fully co-ordinated and integrated response involving a great many agencies and outside support"

The Home Secretary's Review

So why did the Home Secretary ask us to carry out a review of civil emergencies which he set up in April 1988.

It was a combination of the growing number of disasters at the time, the growing media interest, quite properly, the public and parliamentary concern and his responsibilities particularly for the Police and Fire Services to ensure that the response to major emergencies is as effective as we can make it.

We had at that time suggestions covering almost every conceivable possibility for approaching disasters.Some would write to the Home Secretary with technical improvements,minor suggestions, others suggested major structural reforms,disaster squads to be mobilised,centrally based,and even a Secretary of State for Disasters.

But there was no evidence available to us then for the need for a complete root and branch restructuring.The Home Secretary's intention was to identify what gaps existed within our arrangements,and to see what practical improvements,and I stress the word practical,could be made to improve the situation.

The point here is not to change for change's sake but to change where we can be certain that any development will offer a distinct perceptible improvement.We carried out a major consultation exercise asking the views of everyone who we could think with a direct involvement and also taking comments from members of the public and from any other interested body.

This process culminated in a seminar that we held in November 1988 at what was then the Civil Defence College at Easingwold in Yorkshire.A seminar which brought together principally,and this is important,the operational figures,the Chief Constables,the Chief Fire Officers,the County Emergency Planning Officers,the Local Authorities,the Voluntary Bodies, Government Departments,Health and Safety Inspectorates.The people essentially who could say whether it would work.In the course of the consultation exercise and our earlier discussions there was much opportunity for theoretical debate and for suggestions and ideas to come forward.We wanted to explore with the practitioners what from that field could be of benefit.

The Home Secretary announced his conclusions on the basis of our work on 15th of June in a written answer to Parliament and copies of this will be available afterwards.That was supported by a fuller document placed in the libraries of both Houses of Parliament which will be available I suggest for subsequent publication if you wish.

Review Conclusions

The key conclusion the Home Secretary reached was that the prime responsibility for handling a disaster should remain at the local level.Now,that's not solely at the local level,but led at the local level where you have the knowledge of the problem,the resources and the best means available locally to deal with it.Supported,of course,by other agencies and neighbouring forces or services providing assistance under standing arrangements for mutual aid.

I think that conclusion could not have been better tested than by the appalling disasters we faced at the beginning of this year,Lockerbie,Clapham and Kegworth within a space of a couple of weeks.If we had set out to test the validity of the assumption of the belief that the local response is the most effective we could not have devised a more exhaustive test.

In Lockerbie we have the late on Friday evening syndrome in a remote rural area and for good measure the smallest police force in the country with the least resources of it's own.

In Clapham,we have the metropolis and we had the morning rush hour and all the problems that brings in getting resources to the site,getting the injured away and dealing with the consequences.

In Kegworth,interestingly, we had an incident on the borders of three counties.Three police forces,three fire services,three Health Authorities,and all the problems of co-ordination that that brings.

There are,of course, lessons to be learnt and important lessons from all of those events,but I have not heard it said, and certainly not by the services and not by those who are experienced in the area,that any one of those incidents was fundamentally flawed in the locally lead response.

What Lockerbie also showed is just how wide the involvement in a major disaster goes.The number of agencies that were involved in responding at one stage or other to that incident was enormous.It is clear to the Home Secretary that the co-ordination of all the agencies involved,which are properly independent agencies,is essential and that we must have a means of achieving that practical co-ordination to ensure an integrated effective response on the day.

He looked therefore for two things,a way of achieving a national oversight of co-ordinated emergency planning as it became clear in our review, there was no central focus, there was no one point which could grip all the issues.The seminar that we held was unique in the services that it brought together and had it done nothing else,it was valuable I believe,for getting the senior operational figures talking to each other on how best to handle these events,but we could not just leave it as a one off event and clearly we need to ensure effective oversight on a national level.
With that,of course, goes the encouragement and development of effective co-ordination at the local level,at the county base.The picture here varies of course from county to county.In some places better than others but what we would want to see is a system whereby all the services who would be

concerned in the response to a disaster and I here include
not just the accident phase but the longer term follow up
phase which crucially involves the local authorities and the
voluntary bodies to ensure that it is up to the mark as far as
we can possibly make it.

Actions Derived from the Review

To achieve those aims there are two principle vehicles the
Home Secretary has chosen to use.

One is the appointment of a Civil Emergencies Adviser.This
will be a senior figure who will be answerable directly to the
Home Secretary and will advise him and report to him on the
developments he has found and on the progress he has been able
to make in taking forward co-ordinated emergency planning,both
locally and centrally.It is clearly a key post and a
challenging post.The requirement is to command the confidence
and respect,simultaneously,of the emergency services,local
authorities,voluntary bodies,government departments,public
health inspectorate,the media and the public.A challenging
bill that may help to explain why the Home Secretary has yet
to appoint the Adviser.It is hoped that the appointment will
be made shortly.

In addition he has expanded the role of what was then the
Civil Defence College,which is now the Emergency Planning
College,whose principle John Betteridge is with us today.Some
preliminary work has already gone on as to the role the
college can play in addressing incidents which are wholly
peace-time emergencies unrelated to the war time context which
was previously required.Precisely how the College's role will
function here is yet to be determined and will be led very
much and influenced by the views of the Adviser whom the Home
Secretary gave a specific remit to consider the whole question
of training and exercises.I should say in this context that we
very much welcome the establishment of the DPLU and indeed of
other developments at Lancaster University and elsewhere.These
I do not see as competing endeavour in any sense at all.There
is a broad need for a mix of research,for academic work,of
practical sessions,of mixed seminars and I am sure that with
the cooperation and discussion that I know that we will want
to see,and I hope that you would share,that this wide field
can be used effectively and addressed by very different
institutions in various ways.

Of course,a number of people wanted more from the Home
Secretary.It is not the first time that has happened and I
doubt if it will be the last.In particular,there were bids for
legislation for putting a duty on local authorities to address
and plan for civil emergencies.The position here is that the
Home Secretary was not convinced at present that there is a
need for legislation.A great deal of work we discovered is
clearly being done already and the Home Secretary's wish was
to see what we can do now moving ahead with the results we

have got and to leave the door open and take a final view on the question of a statutory duty at a later stage when we have seen how much progress has been made and what need there is in real terms.The door is left open,as he put it,and the position will be kept under review as this work progresses.

Role of the Civil Emergency Advisor

So how will this work go forward ? How will the Adviser function ? Which is the question of most interest I suspect.

Supporting the adviser there is a small Secretariat which I head which is small and only a secretariat,consisting of three of us only.The resources at the Advisers disposal go very much wider than that.Many of you here would I'm sure be among the first to tell a three man Home Office Secretariat that it does not know the answers to these questions and I would agree with you but we do know who does and we know where to go to find the answers.In a sense the resources available to the Adviser in taking forward this work are the entire resources of the emergency services,of local authority associations,voluntary bodies and other organisations that I previously mentioned.It is,of course,to the Adviser to determine his priorities and the methods by which he will work.I preface my remarks now by saying that any thing I say is subject to him overruling it if he so chooses.

But as far as I can see,at this stage,the sorts of routes that might be used will be the discussion of various issues by committees or working groups of those who have the practical expertise.Now in some cases this is going forward already,there is a Department of Health funded working group looking at questions of counselling.The Association of Chief Police Officers are looking at the structure and efficiency of casualty bureaus.The Advisers role here will be to ensure that all the relevant interests have been involved in any working group.Obviously a single service will approach it from its own perspective and rightly the point of having a central focus now to pull it all together is to ensure that the strands that effect other groups and other bodies are brought into play to so that the advice on any one question,like counselling,meets the needs as widely as we can of the entire community affected by this subject.Where,of course, no such body exists presently,then it will be to Adviser to set one up, to commission a research,or to take which ever course he wants to find how best to address the topic.In the longer term I should imagine that this will lead to the publication of Home Office guidance.We already have the Emergency Planning Guidance to Local Authorities,a volume mainly, but not wholly,addressed to civil defence to war time needs.One possible way forward would be to greatly expand the second part of that which deals with civil emergencies in peace time and to provide guidance on all aspects of handling disasters.Guidance which would have been agreed first with senior representatives of the various services. It would carry, I imagine, their mandate, their approval as well as the advisers and it will be supplemented of course by the detailed guidance, manuals, whatever you like, within each service indicating how their staff put it into practice within their own terms. As I say, we are not in the business of telling people how to do their job but of helping people to find out with their colleagues and with whatever help we can provide what is the best way.

The possible areas that may be looked at are, and the list is as long as your arm, the four specifically identified by the Home Secretary in his statement, the question of psychological damage to victims, the efficiency of casualty bureaus, the provision of assistance for disasters occurring overseas and the whole area of training and exercises. Run your mind through more recent disasters and it is not difficult to come up with other needs, communications, handling of the media, involvement and the role of industry. As the Adviser takes up his post and has discussions with senior figures in all those areas I have every confidence that the list will grow considerably longer and will keep my successor as busy as I expect it will keep me.

What power does the Adviser to do any of this ? The answer is in statutory terms - none - , but he has the power of persuasion, the power of public report to the Home Secretary. It would be difficult to explain, I think, to a Public Enquiry why you chose to deviate from a broad norm of guidance unless locally you had a very good reason to do so. We don't want to be in the position of laying down formal rules that must be followed because the situations as we know only too well differ from area to area and the type of event will differ and all we can do is set out broad principles. These will be important broad principles and as I say and will have the backing of the services concerned.

It has been put to me then that the Adviser will be banging peoples heads together. Not an image I really like. Bringing peoples heads together - yes, that will be his skill, his diplomatic skill and the useful part of the knowledge he will acquire from talking to all of those involved.

Conclusion

In conclusion then I would say that each service, emergency service, local authority, voluntary body, whatever, has it's profession pride in the work it does and quite properly but I think of what we have learned above all else over the past two years is that cooperation is the key to success.

The fact that we have come together for this conference is an indication of that and if you look down the list of those attending,the range of interests concerned drives the point home again and again.We can't do it alone,we must talk,we must plan together and on the day we must act together.It is only in that way I believe that we can make the most effective use of the resources and systems that are available to us because, when we are concerned on the day with responding to a disaster,there are no second chances.

THE ROLE OF THE FIRE BRIGADE IN THE PREVENTION AND LIMITATION OF DISASTERS

W. D. C. Cooney, O.B.E., Chief Fire Officer,
Cleveland County Fire Brigade

BACKGROUND

CLEVELAND COUNTY is situated in the North East of England and is bordered by the counties of North Yorkshire and Durham.

The present CLEVELAND COUNTY FIRE BRIGADE was formed in 1974 as a result of the re-organisation of the whole of Local Government within the United Kingdom. The new Brigade combined into one Authority the former Fire Brigades of Teesside, Hartlepool and small parts of the North Riding of Yorkshire and Durham Fire Brigades.

CLEVELAND COUNTY FIRE BRIGADE is divided into two operational divisions which are on either side of the River Tees. In order that the Brigade can discharge its legal obligations, there are 15 strategically placed Fire Stations. The Brigade has a complement of nearly 700 full time professional Firemen and approximately 150 back up staff in the form of administrators, technicians, etc.

By national standards, Cleveland County is a small area covering 58,550 hectares and has a population of approximately 650,000.

On close examination, any visitor cannot help but notice the very high concentration of industrial areas which are dominated, in the main, by chemical and petro-chemical installations. In fact, the majority of the county area, some 13%, is classified as the highest risk categorisation within the Fire Service regulations within the United Kingdom, that is A Risk. Other areas of the county fall within the Special Risk categorisation. This industrialisation means, in effect, that Cleveland County Fire Brigade responds to the largest concentration of chemical and petro-chemical complexes in Western Europe.

The Tees and Hartlepool Port Authority who control the ports within the county are now the third busiest port in throughput, some 40 million tonnes, in the United Kingdom and, in the terms of handling hazardous/dangerous products, have more movements than any other UK port.

Additionally, in the north of the County is located a Nuclear Power Station, centrally a large British Rail marshalling yard and on the western border, a civil Airport. Within the towns, there are high rise developments, shopping arcades with enclosed malls and a number of sports grounds.

There is, without doubt, potential for disaster and emergencies within Cleveland County and the Fire Brigade must ensure that, should an incident occur, we have the capability to effectively and efficiently deal with it. This can only be done by planning and, having planned, carrying out exercises to put the plans to a test. When considering plans, it should be remembered that only when the incident involves fire is the Fire Brigade in control of operations and, furthermore, only those operations which involve the extinction of fire.

INTRODUCTION

From time to time, an emergency incident of disaster proportions occurs in which the Fire Brigade will find itself working alongside, and in close co-operation with, other Services and organisations, some of which may not usually be regarded as "emergency services" in the course of their normal duties. The Fire Brigade's experience in emergency and rescue work and in establishing effective command and control arrangements allow us to function efficiently and effectively as an individual service. It is, however, necessary to recognise, not merely our own specific responsibilities, but to liaise with other organisations who will invariably be involved at an emergency of major disaster proportions.

An incident which, by the nature of the hazards, the number and seriousness of casualties and/or the amount of disruption caused to services, is beyond the capacity of the normal resources of the uniformed Emergency Services to clear up unaided and requires the special mobilisation and co-ordination of the non-uniformed Local Authority services and/or other organisations and agencies.

Cleveland County Co-ordination Scheme for Major Incident

The number of casualties involved should not necessarily be regarded as the main criteria for determining whether an emergency reaches the magnitude of a major disaster. Other factors such as the type and location of the incident, the probable number of services likely to be involved and the time taken in rescuing the trapped and injured and restoring normality, must all be included in the assessment. For example, an air crash on an airport involving 50 casualties who, in a relatively short time, can be extracted and removed to hospital, is not necessarily a major disaster in terms of widespread mobilisation of numerous non-emergency organisations, whereas the same incident in a built-up area could be termed a major disaster, because of the overall disruption to life and to services and the length of time and extensive operations involving numerous organisations required to deal with it. Some examples of incidents which could develop into Major Disasters are:-

a)   Major leakage of toxic chemicals into sewers or watercourses
b)   Major leakage of toxic or flammable gases, particularly when occurring in or near residential areas
c)   Serious accident involving the discharge of radio-active materials
d)   Major fire or explosion
e)   Widespread flooding
f)   Severe damage to a wide area of property due to freak weather conditions
g)   Serious road or rail accidents
h)   Aircraft accidents

Incidents which have occurred at Flixborough and Kings Cross, London - England; Bhopal - India; San Juanico - Mexico; Piper Alpha - North Sea; Chernobyl - USSR; Basle - Switzerland and Seveso - Italy are many and varied and have often involved not only loss of human life but, in some cases, extensive environmental damage.

You may ask, how are disasters prevented? - and the answer must be, with extreme difficulty. However, damage caused by such disasters can be limited, providing such events are planned for. Planning is the key. Having formulated plans, exercising them regularly ensures participants are aware of their role in the event of the plan being brought into effect.

1) PLANNING

1.1 PLAN CONTENT

The plan should cover any type of emergency which would necessitate co-ordination of effort of the Emergency Services and other services, be they Local Authority, Military, Governmental, etc. It should also:-

(a) Make advance arrangements for effective operational control (co-ordination)
(b) Make provision for facilities for use in an emergency
(c) Be comprehensive in content
(d) Be in writing and be simple to read and understand
(e) Ensure it's contents are notified to those responsible for putting it into effect

Most importantly:-

IT MUST BE PRACTICED

1.2 EXERCISING AND REVIEWING THE PLAN

Arrangements should be made to test the plan periodically, preferably by means of a full scale exercise. Furthermore, whenever the plan has been used at a real-life incident, it should be reviewed as soon as possible afterwards and any deficiencies rectified.

There is no better way to test the availability of personnel than to run some form of exercise. In order to ensure that the plans have been drawn up to allow sufficient flexibility, exercises need to be undertaken to ensure that participants are fully in the picture with regard to the possible problems that could arise following a major incident. The method of undertaking exercises are many fold and the size and format can vary enormously. However, when all Exercises undertaken are analysed in detail, they fit into two main categories:-

A)   FULL SCALE EXERCISES

It is necessary on occasions to actually move manpower and equipment to test their availability and capabilities of handling an emergency situation. It is therefore important that major physical exercises are undertaken on a regular basis.

These exercises require to be well planned in advance, with a carefully written scenario and they should be guided towards certain objectives. It is no good having a hap-hazard physical exercise, as no benefit would be gained from hap-hazard results.

Clear supervision and assessment, plus a vital debriefing session, should take place after an exercise and within such a time as that the exercise is still fresh in the minds of the people who undertook it. By this means, maximum benefit from the problems encountered and any mistakes that were seen, shall be gained. It is important that this liaison, and this type of exercise, takes place at a very early stage after the Emergency Plans have been drawn up.

A)   FULL SCALE EXERCISES (Continued)

PROBLEMS ASSOCIATED WITH THE PHYSICAL TYPE EXERCISE

1)   Wholesale evacuation of members of the public
     cannot be undertaken during a physical exercise
     as there is a considerable difficulty in
     evacuating the public in a real life situation.
     It would be even more difficult to move people in
     an exercise situation.   Also, disruption caused
     to the normal way of life within that particular
     community would not justify involvement at that
     level.

2)   The actual use of large amounts of personnel has
     a fairly large cost implication.   Also, it could
     denude available resources away from normal
     activities.

3)   Physical exercises undertaken by the Public
     Emergency Services can only allow limited
     resources of manpower and equipment so as not to
     take away the sharp end service to the public,
     for example fire appliances off normal
     activities, simply to test a plan with a physical
     exercise.

B)   TABLE TOP EXERCISES

     Exercises of this type are used to test the Emergency
     Planning procedures by using table top plans and
     imaginary situations and a carefully detailed, written
     scenario.

B)   TABLE TOP EXERCISES (Continued)

A Table Top exercise is a useful medium to test out management reaction to an emergency situation and their ability to put the adopted plan into operation and provide the necessary information required by participants. It also allows all participants, including the Emergency Services, to evaluate their response.

A simple, straight forward, Table Top Exercise can have considerable advantages in checking for example the plans in respect of advice to public, availability of people in the right places to undertake certain tasks, etc., amending and alteration of the plans as required, without massive disruptions to output.

EXERCISES IN WHATEVER FORM ARE   **REQUIRED**   TO TEST EMERGENCY PLANS

CONSULTATION AND COMMUNICATION IS THE KEY TO THE EFFECTIVE PREPARATION AND IMPLEMENTATION OF PLANS. IT IS FUTILE TO IMAGINE THAT GOOD WORKING RELATIONSHIPS WILL BLOSSOM AND GROW ON THE DAY THE INCIDENT OCCURS. IT TAKES WORK, EFFORT AND PRACTICE (EXERCISES) TO ACHIEVE A LEVEL OF UNDERSTANDING AND COMMUNICATION WHICH WILL PROVE EFFECTIVE IN ENABLING ALL THE VARIOUS SERVICES INVOLVED, TO HANDLE A MAJOR INCIDENT EFFECTIVELY. FORMULATE YOUR PLANS AND DON'T ALLOW THEM TO GATHER DUST ON A SHELF - EXERCISE THEM. AN EFFECTIVE EMERGENCY PLAN WILL NOT PREVENT INCIDENTS OCCURRING BUT, WHEN THEY DO, THEIR EFFECTS, POTENTIALLY DANGEROUS THOUGH THEY MAY BE, CAN BE SUBSTANTIALLY MITIGATED.

References: County of Cleveland Co-ordination Scheme for Major Incidents and Offsite Industrial Emergency Plan(s).

Fire Services Act 1947

AN AMBULANCE SERVICE VIEW OF MAJOR DISASTERS

A. Page, Chief Ambulance Officer,
South Yorkshire Metropolitan Ambulance Service

Preparation

All emergency services must obviously plan for the handling of major
incidents. Some perform this action more vigorously and seriously than
others, but nevertheless, each service has an obligation to the community
it serves to be prepared.

For many years the table top exercise was considered to be all that was
necessary for most organisations to up date and test themselves against.
Unfortunately, in this more sophisticated world, we must now accept as
reality the daily threat that exists for us all.

Added to this is the knowledge and experience of recent years as to the
frequency of disasters, their complexity and the devastation created to
structures not to mention the effects they have on and to people.

Our attitudes and capabilities have changed tremendously over the past ten
years, we no longer view disasters as rare events, instead we watch the
media to see where the latest disaster has occurred. We expect disasters
and therefore training and preparation play a very important and essential
part in our organisational policy and development.

If we consider some of the disasters which have occurred we can see that
they have affected public transport, communities, social life, industry and
sporting events. This proves that disasters have no respect for societies
and therefore no amount of planning will remove the need for sufficient
manpower establishments being maintained or appropriate equipment being
purchased or sufficient training being undertaken.

It is true that the public sector organisation including the Police, Fire
and Ambulance Services are constantly being reviewed and having their
budgets and manning levels scrutinised. This is correct if all of the

organisations are allowed to develop to meet public requirements for safety, if and when necessary, but all too often the fact of the matter is that we have reductions imposed on our services with the expectation that we will still be capable of assisting society in the event of a disaster. Even to the inexperienced and less knowledgeable amongst us this is not just unfair to assume it is positively dangerous to life. It also effects the morale and safety of the rescuer who places him or herself at risk with a clear understanding of the dangers and support available to him or her.

Successful disaster planning and handling is not about providing large numbers of personnel, vehicles and equipment. It is about providing well and adequately trained and equipped personnel and vehicles to deal with the needs of the disaster and the injured or trapped. There is no use in any service flooding a scene with valuable resources which will not be positively used. To send too much manpower into any incident is poor resource management and usually leads to confusion, lack of coordination and criticism by all concerned.

How Do We Know What is Required?

The answer has to be related to having, in the first instance, a good sound knowledge of our area of responsibility, its hazards or potential hazards. Keeping updated with the changing environment and geography is absolutely essential. The successful response and management of an incident depends on many things.

Adequate Planning

Regular updating of plans, knowing your problem areas and/or establishments. Recording hazards and quantities and most importantly keeping up to date with change and developments. Developing positive liaison with local industry and safety officers, as well as other emergency services, local authorities and voluntary organisations.

Adequate Inter-Service Communication

Meetings should be arranged between service personnel of all ranks in order
for each group of staff to understand the role of other services.
Equipment, training and ideas should be seen and discussed by all in order
to assist in operational situations. Face to face meetings develop
positive understanding and relationships.

Effective Staff Training

Regular training courses and exercises where staff are pushed to the limit
of their training and service capability and knowledge. New techniques and
ideas should be encouraged and developed and then shared with other
organisations. We should never be too proud or arrogant to think we can
handle any situation we may face. That is not true and seeking the
opinions of others will save lives at a later stage, as well as being more
cost effective.

Understanding Service Roles

Each emergency service must understand the role and capability of the
other. No one service can cope solely with an incident and no one service
is in total charge from beginning to end. Specialist roles will be
required at various stages and senior officers have to define these points
during training sessions and briefing meetings.

Acting Swiftly

The Police, Fire and Ambulance Services within the United Kingdom are now
extremely well informed of the effects of disasters and the need for swift
action and response. These actions must be performed in a professional,
calm and positive manner. There is little doubt of the danger to the
successful management of any incident if panic is allowed to creep in.
Bureaucracy and in some cases, the use of computers may cause delay in
responding and therefore I believe policies should only be implemented and

systems installed when it has been proven that the common sense of the operator or service plan is being enhanced not impeded.

Provision of Adequate Emergency Cover

Chief Constables, Chief Fire Officers and Chief Ambulance Officers must never accept reductions in manpower without voicing a positive opinion. If they agree then so be it, after all they are the officers who will be held to be totally accountable during the event and aftermath. I do not agree with service growth which is unnecessary. However, senior officers must ensure that reduction or proposed reductions in their service manpower or resources which will or may effect their service capability in daily emergency responses or disaster incidents are positively and effectively recorded. This action ensures accountability in the right quarters, should an enquiry be instigated at a later stage.

Ignorance or blind acceptance is no defence for any "Chief Officer". Abdication of ones role, responsibilities and professionalism for a quieter life will result in poor service morale and lack of respect and eventually the distruction of that particular service and its capabilities.

Decision Making

Senior officers of the three emergency services of the Police, Fire and Ambulance Service must be capable of making positive decisions in respect of their professional role at any incident.

This requires these people to be in the possession of leadership qualities, to be alert and to be well trained in their role. I must admit that the Fire Service have been of great value to me in developing this role amongst my officers. Responsible officers must ensure their involvement and positive liaison between senior officers of all involved services and organisations in order for an effective team response to any large scale incident.

You must accept responsibility and not rely on others to do your job. If you abdicate your role you must expect to be highly criticised or even removed from your post.

Foresight in Planning and Training

Foresight is an essential part of planning. It is no excuse to say "it will never happen to me". During my career as Chief Ambulance Officer in South Yorkshire, I have had to deal with the Orgreave Riots of 1984 and the associated pit village disturbance following the NUM dispute with the National Coal Board. These lasted for six months and required my ambulance personnel to respond in protection helmets, fireproof suits and protected vehicles.

Had we not watched and taken seriously the build up of that dispute, we would not have been prepared to perform our role. We became the first Ambulance Service in the United Kingdom to wear "riot gear". I must admit that this was not popular at the time with certain groups of Politicians, but for me as the Chief Ambulance Officer it was the only way my staff could do their job in relative safety and that was the most important thing.

This year in April (1989) my service were again called upon to respond to a major disaster - the "Hillsborough Football Stadium". When Liverpool Football Club and Nottingham Forest Football Club were playing. I will not go into any major detail, but it is sufficient to say that 95 people lost their lives that day and 172 were injured.

My service responded with 41 ambulances and a total of 120 Ambulance Service personnel on scene with another 20 involved with the hospitals and control centre. Had I not taken notice of the information supplied after the Bradford City Football Club Stadium Fire, my service would not have been so prepared or so well trained and equipped to effectively deal with the disaster. These two incidents are of course larger than the usual serious incidents which take place every day in this country, but they do prove my point in taking note of situations and having foresight to plan for anything which may happen. It is too late to plan after the event.

What About Future Disaster Prevention and the Ambulance Service?

The Lord Justice Taylor enquiry into the Hillsborough Disaster was supplied with a list of recommendations from my service with regard the role of the Ambulance Services at Sports Stadiums where large crowds were expected, for example pop concerts. I am pleased to say that the interim report issued by the Right Honourable Lord Justice Taylor on the 4th August 1989 has given the Ambulance Service strength it has never had before. More importantly, it places responsibility on organisers, Licensing Authorities and the Police to seek guidance from the professional Ambulance Services on matters related to crowd safety, treatment and safety certificates.

So in future we will have a major role to play in crowd and stadium safety, but we must not forget the developments taking place within the Ambulance Service which increase our response and effectiveness to major incidents.

My service like many others now have paramedics who can set up fluid drips, give drugs, monitor heart rhythm and if necessary defibrillate the heart back into rhythm and intubate unconscious patients by inserting an airway past the tongue and down to the lungs.

We are trained in civil disorder and footsquad activities as well as being trained in ambulance vehicle responses to riots and large crowd problems. This in itself requires many hours of specialised training on the scene and with operational management and also in updating and coordination.

The modern Ambulance Service is just 41 years old and developing so fast that we must start letting others know of our capabilities.

THE ROLE OF THE POLICE AT THE SCENE OF MAJOR INCIDENTS

J. Ashman, Superintendent, Northumbrian Police
Northumbrian Police Headquarters

(Editorial Note:-Mr. Ashman agreed to prepare and present this paper at very short notice and the Conference Organisers are very grateful indeed for his participation and cooperation)

The primary responsibilites of the Police now,at this moment in time are the protection of life and property,the prevention and detection of crime and the preservation of the Queen's peace.

That is a very wide range of responsibilities and that does not change when we come to the scene of major incidents.The role of the Police Service at major incidents is very much a co-ordinating role.

I would not dream of telling the Fire Service how to put out fires,I would not dream of telling the Ambulance Service how to save life or tell the Gas Board how to deal with a gas leak.But when it comes to a major incident and you have all of these services involved,plus the Press,volunteers and members of the general public then someone has to co-ordinate,someone has to put it all together to allow the specialist to get on with their job and that role is very much the role of the Police Service.To co-ordinate all of the actions to ensure that at the end of the day all goes well.

When we talk of major disasters,there are two phases to major disasters.There is the initial impact and life saving phase which may last hours or may last nearly a day.The second phase is the back up phase which may last days,weeks or even months and sometimes there there are split responsiblities.Let me go through some of the things that may happen.

As the Police Service is out and about twenty four hours every day it is more than likely,not always but more often than not,we will be first at the scene or be first to recieve notification of the incident.Therefore one of our first responsibilities is to ensure that everybody required is called to the scene.Obviously,initially Police,Fire,Ambulance Then that is built up dependent on the type of incident it is,is it a gas explosion, is it a rail accident,is it the Civil Aviation Authority,as to who else we call.So that is a very important initial function for the Police,the Call Out.

Then as further information is required,the Police Officer at the scene acts as a Situation Officer who supplies that information and based on that the decisions are made as to else is called to the scene.You have seen on television so often the problem of traffic,so immediately a traffic control plan has to go in to protect the immediate area of the scene in what is called an inner cordon and then further away diversions to direct other people away from the surrounding area.But there are people who have to get into the area,people live there,people work there and therefore a system of roadblocks has to be set up and that is the responsibility of the Police Force.In particular to ensure that there is an efficient one way system for the emergency services.The question has already been raised about ambulances getting into places and that is our responsibiltiy to ensure that they do get in and do get out and if this means knocking down some bollards,then so be it.

When people have arrived at this impact phase the biggest problem is trying to bring order to disorder and therefore site management is a great responsibility,the Fire Service is fighting the fire and trying to save lives and the Ambulance Service are trying to do things then there is tremendous problems of site management and if you do not manage it then it becomes chaos.This is the responsibiltiy of the Police officer at the site.

After this phase we start to look at liasion with the Local Authority and the resources of the County Emergency Planning Officer.He co-ordinates the Local Authority's response so that the police are not contacting the Education Department of the Local Authority asking for schools for feeding centres or as temporary mortuary use.This would be done by the CEPO.

Obviously there are going to be casualties and peolpe are going to want information.This is where the Casualty Beaurau comes in and this is our responsibility.To give this information out to the public and the press.Again,the handling of the press is our responsibility.To prevent the pirate type press man getting in to get his photograph of the scene and to hold press conferences and to ensure that the press are give a fair crack of the whip.On the question of the press i would simply say that,and this is my personal opinion,that we should educate the public and the press that when that magic number comes onto the screen that not every uncle or cousin should ring and block the system.Perhaps we should educate television presentors to say that only one person per family should ring and this might cut down the number of calls.

As we are going to have fatalities,it our responsibility via the Coroners Office and Coroners Officers to set up temporay mortuaries, identification of the dead,permanent mortuaries and dealing with the next of kin.In any disaster there is going to be a question of property,property of people killed or injured in the disaster,in an aircraft accident of piecing it all together,securing property for the investigation people,all of this needs to be secured and looked after by the Police Service.

Then later we come to the investigation of the cause.Was it terrorist activity as at Lockerbie,or was it an accident and is there a public enquiry as at Hillsborough.We have to tie all of these things together and although some peoples part may be over within days,for example it may take three months work to indentify a body from a fire.The Police Service has long broad responsibilities and it is the co-ordinating of all of these things together that is the function of the police.

Through the Assoication of Chief Police officers we have a sub committee the chairman of which is responsibile throughout the country for investigating the proceedures used at major disasters.After each disaster the report by the local police is circulated nationally and any recommendations are evaluated and if necessary included into our own contingency plan.

THE ROLE OF THE COUNTY EMERGENCY PLANNING OFFICER

J. R. Parker, Civil Protection Office,
Cambridgeshire County Council

We can all recognize the dichotomy that currently exists in the present role of the County Emergency Planning Officer.

On the one hand his principal commitment is undoubtedly in respect of making plans for the facilities and services as set out in Schedule 2 of the Civil Defence (General Local Authority Functions) Regulations of 1983 - this has been reinforced by the Civil Protection in Peacetime Act of 1986 which enables him to utilize the Local Authorities Civil Defence Resources in response to peacetime emergencies. On the other hand now, the County Emergency Planning Officer is obliged to respond to requests from Industry - with some help from the Health and Safety Executive - to produce Emergency Off-site Plans to comply with the Control of Industrial Major Accident Hazards (CIMAH) Regulations of 1984.

Some industries are now actually requesting off-site emergency planning assistance despite their not coming within the CIMAH Regulations. In Cambridgeshire, we respond positively to such requests and assume the responsibility for the production of an off-site emergency plan. All aspects of consultation, coordination and public information are controlled by an Emergency Planning Officer. Companies are prepared to pay for this service - and why shouldn't they?

The County Emergency Planning Officer who has logically developed and written his Emergency Plans as required by the 1983 Regulations will know that there are many common features that are complimentary between his Wartime plans and those for Peacetime emergencies. The astute County Emergency Planning Officer will recognize this at a very early stage and will turn it to his advantage. Since 1981, some County Emergency Planning Officers have been applying this 'All Hazards' approach to emergency planning with good effect.

So, first and foremost - part of the County Emergency Planning Officer's role is that of Plan writer. He must prepare 'His County'. He must do his homework - gather his Resources (use a database, if possible) - he must do his preparation and research. It may seem glib to say but he must also win the support of his Councillors and Chief Officers alike. He must, therefore, be a persistent salesman and a first class diplomat.

---

* Cambridgeshire Emergency Planning Officers are called 'Civil Protection Officers'.

Having produced a plan or plans he must educate and train the people who are going to 'work' the plan(s). This is easier said than done. Local Government Officers have little time to spare. Training too - is time consuming - but essential. I, personally, recommend small doses at regular intervals.

Next, the plan must be exercised from top to bottom - another time consuming effort - but essential. The plan must be 'proved'. Modifications to the plan must be expected and accepted - the perfect plan has yet to be written. The County Emergency Planning Officer is a very busy man indeed.

In the event of a major peacetime disaster - the County Emergency Planning Officer's role is pivotal and primarily an advisory one. His Local Authority has two main objectives to achieve in an emergency; first, it must maintain its own services and help people in distress and secondly, to coordinate the work that is being done by the various organizations that are giving help.

The County Emergency Planning Officer's role is continually evolving and will become even more prominent in the future. He must be constantly liaising with the Emergency Services, Voluntary Services, Utilities, Public and other bodies. He must, above all be a convincing salesman confident in his product. If he is to be seen as an adviser to the Chief Officers, then he must be trusted and believed.

Local Authority resources are unlikely to be able to meet all the needs in a major emergency, particularly one that is widespread and lasts for more than a few days. In Cambridgeshire, at neighbourneed level, we prepare for such disasters through a scheme called Emergency Planning in the Community: EPIC. This involves volunteers drawing up a plan, from a model supplied by the County Council's Civil Protection Office. The intention of this plan is to make the best use of Community Resources in a major emergency - whether they are used in support of the Local Authorities or alone, should a community be cut off. In Cambridgeshire, the County Civil Protection Officer has recognized the valuable contribution that can be made by the Volunteers and what is more their contribution takes emergency planning to grass roots level, from County through District to Parish and Ward. Volunteers undertake training in a range of skills, useful if a disaster occurs locally. These include Rescue and Firefighting, First Aid and Casualty Handling, Communications (RAYNET, REACT and Emergency Centre Comms.), and Rest Centre Care. The latter involves Emergency Feeding and Accommodation and the collection and distribution of Information on victims of disasters.

To sum up, finally, the Role of the County Emergency Planning Officer with regard to a Peacetime disaster is to ensure that his plans and preparations exist, that they are workable, his people trained, and that his position is ultimately that of a well-briefed adviser.

CHEMICAL DISASTERS AND THEIR LESSONS

V. C. Marshall, Honorary Fellow,
Department of Industrial Technology,
University of Bradford

INTRODUCTION.

The purpose of this paper is to examine the nature of chemical disasters as a means of determining the nature of the emergency measures which are likely to be required in the aftermath of such disasters. It aims, among other things, to give indicative ranges for the harms which may arise.

The question as to what constitutes a disaster has to be faced at the outset. The writer has elsewhere characterised the number of fatalities an accident causes as the "default" criterion of a disaster [1]. That is to say that where there are substantial numbers of fatalities, say ten or more, resulting from an event, then the event is clearly a disaster. However other events, not involving substantial numbers of fatalities, may still be regarded as disasters if there are other consequences of a serious nature. Such factors could include, as examples, severe property damage, widespread dislocation of personal lives, or long term environmental effects.

There are a number of ways of defining chemical disasters and the definition given by the author in this paper is only one of them. They are defined here as:— " Chemical disasters are acute events which produce, on a large scale, and through such agencies as chemical fires, explosions, or toxic or corrosive releases a number of harmful effects. These effects include harm to the health and safety of employees or of the public, or damage to property or to the environment. Though the events themselves are acute, the effects they produce may be, in certain cases, chronic or long term". A list of the possible harms is given in Table 1.

The events under discussion broadly correspond to the "major accident hazards" of the European Directive [2] and the CIMAH (Control of Industrial Major Accident Hazards) Regulations [3] which implemented the Directive in the United Kingdom. However, though these Regulations apply only to fixed installations and thus do not apply to the transport of hazardous materials, this distinction will not be made here. Nor will the discussion be applicable solely to those incidents which arise from installations which are fall under the provisions of the CIMAH Regulations.

WHAT GOES WRONG IN A CHEMICAL DISASTER ?.

A chemical disaster results from the large scale, unplanned, release of energy or matter or some combination of the two. The principal forms of such releases are fires, explosions and toxic or corrosive releases. There are other forms of chemical hazard which are discussed in Chapter 16 of the author's "Major Chemical Hazards" [4]. It may be said that these other forms may give rise to death or injury to employees and to on-site buildings but are unlikely to produce significant off-site effects.

In general the releases under discussion represent a sudden loss of containment as for example through the failure of a pressurised system or through some form of mal-operation.

Energy may be released without significant dispersion of matter as in the case of the detonation of dense explosives. Alternatively matter may be released without significant release of energy, such as in the spillage of volatile liquids, but the vapour arising from these may be dispersed by externally supplied energy from the wind.

Fires and some explosions release chemical energy . In the particular form of a vapour cloud fire chemical energy is released but such energy release is preceded by a large scale dispersion of matter in the form of a cloud of flammable vapour. Such dispersion, when it arises from the spillage of liquefied gases, involves an initial phase in which the predominant form of energy is provided by the internal energy of the system. In the second phase gravitational energy and then energy provided by the wind predominates.

There have been many cases in which the loss of containment of liquefied flammable gases has led to a cataclysmic fire. Depending upon circumstances such releases can give rise to a ground level flash fire or in some cases to a fire ball. Some fire balls have been characterised by a burn up rate of the order of 1 tonne per second. [This is comparable with the overall rate at which gas is being extracted from the North Sea.]

A clasical example is that of the Flixborough explosion of 1/6/74 where a release of ca 40 tonnes of cyclohexane vapour led to an explosion with an blast energy release equivalent to a ground level explosion of 32 tonnes of TNT.

Toxic release takes a number of forms. In general it may be said that though the effects of explosions and fire are basically similar regardless of the chemical nature of the substances which give rise to them the effect of a toxic substance is unique to that substance. Toxics can differ in their mode of entry into the body; some enter through the stomach, some enter through the skin and others through the lungs. They differ in the organs which they attack. Some damage the respiratory system, others the central nervous system and others damage the metabolism. Corrosive substances are a special variety of toxic substances which severely damage body tissue and vegetable matter.

SOME ASPECTS OF CHEMICAL FIRES.

Chemical fires may be distiguished from the fires usually encountered by the fact that they principally involve the combustion of liquids and gases rather than solids. They are distinguished by their intensity and by the speed with which they spread. Some solids fall into this category by being capable of giving rise to anaerobic combustion because of their chemical composition. Cellulose nitrate is a substance of this character.

Fires of the sort described have been known to give rise to death and injury outside of the site of an installation as for example in the Cleveland, Ohio, fire of 20/10/44 and in the Mexico City fire of 19/11/84. In both of these cases it might be argued that the circumstances were not typical of those experienced in the United Kigdom today as in both of the cases cited housing had been permitted very close to the site boundary which, in its turn, was very close to the point at which containment was lost.

Perhaps more germane to the discussion is the propylene tanker which disintegrated at San Carlos, Spain on 11/7/78. In this case some 23 tonnes of propylene were released and the resulting conflagration killed 215 people on a campsite. Though the risk of such an event in the U.K. is very low nevertheless so long as liquefied gases are transported through populated areas the risk is not zero.

Detailed descriptions of these, and similar, incidents are provided in Chapter 9 of Ref [4].

A further problem which has recently come to attention has been the consequences arising from the run off of fire fighting water when this carries with it materials harmful to water supplies. In the author's view the problem may have been with us for a long time but that it is only recently, with growing awareness of environmental problems that it has received media and hence public attention. For example the Flixborough explosion referred top above released considerable quantities of chemicals into the River Humber but this attracted little attention when compared with the attention given to other aspects of this disaster.

Three incidents of this sort which have occurred in the last seven years have come to the author's attention. These are the incidents at Morley, near Leeds (13/2/82) where some tens of tonnes of paraquat and diquat were carried by fire fighting water into a river. The Basle incident of 30/10/86 has been well publicised. Here, following a fire, some hundreds of tonnes of miscellaneous pesticides including 12 tonnes of an organic mercury fungicide entered the River Rhine. The author has calculated that the fire fighting water used would have flooded a football pitch to the depth of around one and a half metres. The most recent example is the incident at Chateau-Renault, France on 8/6/98. Here an explosion was followed by a fire and the run-off led to pollution of a number of rivers.

ASPECTS OF EXPLOSIONS.

Physical explosions from the loss of containment of liquefied gases have been known since the Industrial Revolution. Originally the liquefied gas was steam from boilers and considerable, though localised, death, injury and damage was caused by such explosions. Today a major contibution to such physical explosions comes from the loss of containment of liquefied gases such as LPG. Usually where flammable gases are involved the resulting fire/ vapour cloud explosion causes so much damage as to divert attention from the initial physical explosion.

Chemical explosions may take place in gases and liquids. Such explosions may be termed dense explosions. There is a vast amount of information on the effects of these because of their military and commercial importance. The generalisation of Hopkinson in 1915 that for varying masses of explosive of the same chemical composition a given level of effect will occur at the same scaled distance is still valid. This law states that the radius at which a given effect will occur is proportional to the mass of explosive to the power one third. Thus increasing the mass of explosive by one thousand times will give the same effect at ten times the radius.

Fig 1 gives scaled distances for various indices of damage to houses A Class being total demolition, B class being damage so severe as to require demolition and Cb to be just worth repairing. The value of scaled distance for 50% mortality the present author calculates as being a scaled distance of ca 18.

Chemical explosions in vapour/air mixtures may yield broadly similar results from the point of view of disaster preparedness. The TNT equivalence may be as high as 0.8 times the mass of flammable vapour if allowance be made that when TNT explodes at ground level approximately half of its energy is dissipated in forming a crater. Vapour cloud explosions do not produce craters.

In the Flixborough explosion all of the 28 fatalities occurred within a radius of 125 metres from the epicentre of the explosion. There were no survivors within this radius. 36 employees were given hospital treatment for injuries and 52 members of the public were so treated. Hundreds were given first aid.

The damage pattern was unsymetrical but in a northerly direction 80% or more of houses were damaged up to 4 kilometres from the epicentre with light scattered damage up to 8 kilometres from the epicentre.

Flammable dust/air explosions are locally devastating but do not approach major vapour cloud explosions in the scale of their effects. They do not occur in the open air and are confined to buildings.

TOXICS.

The range of injury from toxics depends upon many factors. These include the quantity and inherent toxicity of the substance released, the strength and direction of the wind and the stability of the atmosphere. Computer programmes are commercially available which can model such releases. Ref [5] gives indicative figures for a factory using bulk chlorine that there would be a 10 in 1 million chance of death at 350 metres and 0.3 in a million years at 1200 metres. Similar figures for ammonia would be 150 metres and 500 metres.

THE SEVESO DISASTER.

This is an important example of a disaster in which there were no fatalities. Yet the event at Seveso was considered to be so serious that it was responsible for the whole question of major accident hazards being referred to the EEC which eventually resulted in the European Directive.

Here a run-away reaction released a plume of highly corrosive chemicals which also contained dioxin which is a virulent chlorachnegen. 447 people were treated for causic burns of whom 34 were subsequently diagnosed as suffering from chlorachne. A total of 179 cases of chlorachne were eventually diagonosed the majority of them being mild cases. 3000 small animals and 12 large animals died of caustic burns.

There was widespread environmental damage over 4 sq km and there 737 long term evacuations of people from the contaminated area.

PREVENTION AND MITIGATION.

Control of chemical disasters falls into two categories, prevention and mitigation. In the EEC all installations which fall within the provisions of the EEC Directive on Major Accident Hazards are required to satisfy their competent authorities that adequate provision is being made to prevent such accidents from occurring. This is done in the UK through the medium of a "Safety Case" [6]. The installation is require to draw up an on-site plan to deal with emergencies and to cooperate with the appropriate off-site authorities in drawing up a plan to deal with any off-site consequences. Such plans are only likely to be effective if tested by regular exercises.

However there is no legal requirement of the EEC or of the UK that chemical producers who fall outside of the provisions of the Directive should draw up such plans or conduct such exercises. Some do so voluntarily but all should do so in their own as well as the public interest. Nor is there any such provision arising from the Directive to cover transport. A number of operators, as for example the chlorine producers, have such plans and cooperate in dealing with emergencies. The Hazchem labelling system has been a major step forward in this area.

SOME FURTHER LESSONS.

The major lesson is that control has to be strategic rather than merely tactical and this means that control must start in the Board Room. In a chemical process this also means that safety starts at the beginning in the choice of the process and the key decisions on how it is to be implemented. In the past in many cases the process was designed and the plant was built before any serious consideration was given to the question of safety. the first step in this direction was sometimes to advertise for a Safety Officer. No one in the UK would do this today but there are many plants in existence which originated in this way a couple of decades or more ago. In these installations people are struggling to overcome strategic deficiencies by tactical measures.

A major strategical consideration is that of inventory. As Trevor Kletz might put it "What you don't have can't catch fire or poison people" [7]. The CIMAH Regulations are inventory based and have led to widespread inventory reductions. These are generally sound provided that they do not result in over-frequent deliveries or collections which carry their own hazards in their train.

A second major strategic consideration is that of site layout. Once the most hazardous areas have been identified their population should be reduced to the minimum consistent with safe operation. It is folly to accommodate the training office in the control room. Flixborough was a glaring example of this; the explosion reduced not only the control room but the main office block to rubble. If it had not been Saturday afternoon the death toll there might have been of the order of one hundred and fifty to two hundred.

Finally every attempt must be made to learn the lessons of the past. Every calamity needs to be closely analysed to derive lessons for the future.

REFERENCES.

[1] Marshall V.C. "A perspective view of industrial disasters". ATOM Feb 1988 UKAEA, London.

[2] European Community Directive "On the major accident hazards of certain industrial activities". 82/501/EEC Official Journal of the European Communities No L203/1 1982.

[3] Health and Safety Executive "A guide to the Control of Major Accident Hazards" HMSO 1985.

[4] Marshall V.C. "Major chemical hazards" Ellis Horwood Ltd 1987.

[5] Health and Safety Executive "Risk criteria for land use planning in the vicinity of major industrial hazards".

[6] Lees F.P. and Ang M.L. (Editors) "Safety cases" Butterworths 1989.

[7] Kletz T.A. "Cheaper ,safer plants or wealth and safety at work" I.Chem.E. Rugby 1984.

TABLE 1. CATEGORIES OF HARM.

(1) BODILY INJURY.

Fatal injury
Disabling physical injury
Non-disabling physical injury
Mental injury*
Fatal disease
Non-fatal disease.

(2) SOCIAL IMPACT.

Evacuation
Loss of livelihood.
Production of social trauma.

(3) ENVIRONMENTAL DAMAGE.

Loss of amenity for residents.
Injury to flora and fauna.
Damage to water supplies.
Long term ecological damage.

(4) PROPERTY DAMAGE.

Damage to equipment, stored materials and on-site buildings
Damage to property in the public domain.

(5) FINANCIAL LOSSES

Losses to the company.

Compensation for death and injury.
Loss of share values.
Loss of sales and customers.
Loss of public confidence in company
Loss of public confidence in the industry

Losses to the public.

Costs of combatting emergencies
Costs of recovering bodies
Costs of conducting inquiries
Loss of property values in public domain

Notes on table 1.

* may be sustained without the victim suffering physical injury
and may affect survivors, the relatives and friends of direct
victims, members of the emergency services and spectators.

Some of the items of cost listed above may be insurable but many
are not.

FIGURE 1

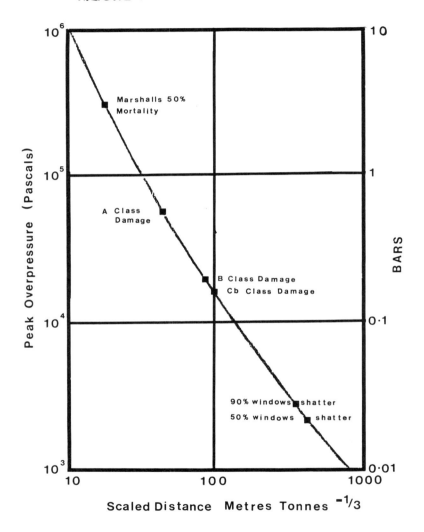

Overpressure v Scaled Distance

WATER POLLUTION —
A HAZARD TO COME?

H. C. Wilson,
Postgraduate School for Studies in Industrial Technology,
University of Bradford

INTRODUCTION

A.Z.Keller (1) defined a "disaster" as

"when ten or more fatalities result from one event over
a relatively short period of time."

Massive water pollution incidents have not produced large
scale fatalities within the United Kingdom within recent
times.The last recorded major event to produce large scale
fatalities was the cholera outbreak in London in the mid
1800's and was due to the contamination of drinking water
supplies by sewage.The cause was remedied by Dr.John Snow who
removed the handle from a water pump in Soho that was drawing
water from a contaminated spring.The number of people who died
was in the thousands,i.e. a disaster of magnitude 3.

Though the time scale if death from water pollution occurs
is longer than that of a typical disaster such Piper Alpha or
Clapham Common,which are measured in a few days,one can still,
with a little elasticity,interpret it as a relatively short
period.

Whilst to the people of Camelford in North Devon the release
of very high levels of aluminium sulphate into their drinking
water supply was undoubtedly a "disaster".The effects of
burned mouths,blistering of the skin and severe abdominal
upsets were proof sufficient.However,the long term effects of
ingestion of high levels of aluminium for a short period are
unknown.The effects of consuming drinking water containing
concentrations of aluminium above the recommended level for
long periods and the link with Alzhiemer's Disease are well
accepted.As has been pointed out,the classification of
disasters may at a future date have to be modified to take
account of long term effects.

For the purposes of this presentation a pollutant is defined
as;

"any matter,substance or organism which upon ingestion by
man will produce an adverse effect"

Irregular transient pollution incidents affecting water
courses increase year upon year.Within the United
Kingdom,nationally,60 - 70% of our drinking water is
abstracted from surface water sources.The existence of the
potential for a disaster is therefore high.

THE SCALE and NATURE of THE PROBLEM.

Irregular Pollution Incidents

Since the formation of the ten regional Water Authorities in 1974 the collation of records, in the main, regarding water pollution incidents has been considerably improved, although some Authorities did not start to keeping detailed records until the mid-eighties.

During 1988 over 23,000 transient pollution incidents occurred from industrial, agricultural, sewage and sewerage and other miscellaneous sources. These transient incidents did not include fertiliser and pesticide leaching which is seriously affecting the quality of water and is occupying much of the attention of the media at present.

The number of incidents per year are given in Figure 1 inspection of which shows that the incident rates were reasonably constant prior to the period of 1982. After this date there is a dramatic increase in the annual numbers of incidents and that rate of increase is still continuing. A possible reasons for this increase is given in Wilson (2).

The percentage of incidents which occurred in 1988 from the four groups detailed above are shown in Figure 2. Industrially related incidents account for 37% of the total and represents a drop of 2% over the figures for 1985 whereas the farm and sewage sources show increases on previous years.

Incidents are currently occurring at a rate corresponding to a doubling time of six years.

The group labelled "Others" contain pollution incidents arising from natural sources and in general these are not harmful to man and will not be considered further.

Assuming that as a best estimate that 10% of pollution incidents could be regarded as "serious", i.e. would require action to prevent contamination of the water supply, and assuming that there are approximately 400 major water treatment plants, then, each plant could be subject to approximately 6 major threats on average per year.

TOTAL POLLUTION INCIDENTS
ENGLAND & WALES

ACTUAL INCIDENTS TREND LINE

Number of Incidents

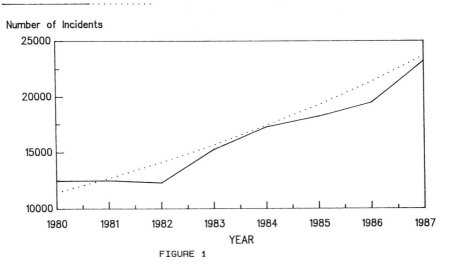

YEAR

FIGURE 1

TOTAL POLLUTION INCIDENTS   (1987)
ENGLAND & WALES

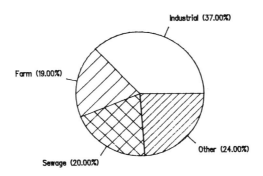

FIGURE 2

NATURE and SOURCE OF POLLUTANT SUBSTANCES

Industrial.

The major pollutants from industrial sites are either chemical or oil spillages.Within the U.K. there are over 50,000 different chemicals either in use or being manufactured(3).Many of these chemicals are toxic,corrosive or mutagenic,i.e.capable of producing DNA mutations within mammalian cells.

Agricultural.

Agriculture represents four possible sources of pollution,

> a).Animal slurries
> b).Release of "cides"
> c).Oil spillages
> d).Silage effluent

Animal slurry is a fact of life for farmers but for the Water Authorities the accidental release of slurry is becoming increasingly problematical to contend with.Animal slurry is rich in enteriobacteria many of which are harmful to humans if ingested.The levels of these bacterium attainable in an abstracted water supply after a release of slurry can easily exceed the effectiveness of the disinfection procedures presently employed in water treatment and pass through unaffected.

"Cides",as defined above, is a generic expression used to cover the many insecticides,fungicides,herbicides and pesticides used in modern farming methods.These "cides" are released into river during spraying,sheep dipping or by accidental spillage.Several of the "cides" are used to maintain the quality of the silage during storage and will be released during an effluent spillage. Many of these substances are extremely toxic to man in very small doses and some are again mutagenic.

Silage effluent contains extremely high levels of biodegradable organic matter and if released into a river will quickly completely eliminate the oxygen content of the water to zero.Silage effluent is highly coloured and can be tasted in water at very low concentrations.

Sewage and Sewerage

There are three main causes of release of raw untreated sewage into a river course;

a).Storm over flow at a treatment works,
b).Breakage of a transit pipe
c).Seepage from a septic tank

The raw untreated sewage contains many pathoses to man and the levels of these bacteria attained after a release of sewage will exceed the disinfection system employed in water treatment systems.As well as the enteriobacteria many viruses are excreted by man via his waste products,e.g.poliomyelitis,which are immune to chlorine disinfection.

A summary of this section is presented in Table 1,

TABLE 1

Effects of Pollutant

	Toxic	Corrosive	Mutagenic	Infective
Chemical	+	+	+	−
Oil	+	−	?	−
Animal Slurry	−	−	?	+
"Cides"	+	−	+	−
Silage Effluent	−	−	+	+
Sewage	−	−	+	+

Several water soluble components of petroleum products upon chlorination are capable of inducing cancerous growth in mammalian cells.
There is a growing belief amongst scientists that some of the enterioviruses may have mutagenic effects.

POLLUTANT DETECTION SYSTEMS.

The Water Authorities rely on various methods to detect transient pollutant slugs,these are;

> a).Continuous On-line monitoring of Physical Parameters
> b).Continuous fish monitoring
> c).Discrete sample analysis
> d).Reportage

a) On-line monitoring

On-line monitors are placed ahead of the abstraction point and the information relayed by telemetry to a central control station.The parameters that are routinely monitored are pH,conductivity,turbidity,oxygen content and temperature.

b).Fish monitors

Several species of fish are routinely used in continuous flow devices.The most commonly used species are brown and rainbow trout.The monitors are based on the theory that fish respond to the level of pollution by changes in their activity pattern.These changes can be compared mathematically to the normal activity pattern and when a preset level is exceeded an alarm is triggered.

c).Discrete Sample Analysis

Water Authority employees collect samples of river water below known hazardous sites,or from areas of the river below known fish kills.These samples are taken to the nearest analytical centre where they are screened for the presence of pollutants.

d).Reportage.

The reporting of chemical,sewage spillages or other forms of pollution incident by the originator,the reporting of fish kills by the general public,the emergency services or Water Authority employees have been abstracted from the Annual Reports of the Water Authorities and are presented below in Table 2;

TABLE 2

Source of Information

Reporter	Percentage of Total
Public	53
Emergency Services	3
Offending Party	4
W.A.Employees	27
Others	13

Hence,the Water Authorities rely heavily on the goodwill of the members of the public for their information with regard to pollution incident reporting.It is worthwhile noting the very low response from originators of the pollution incident.

There are,however,disadvantages associated with each of the above methods when applied to the detection of transient pollution incidents.
The on-line monitors may respond if the pollutant alters the pH or conductivity of the water,but,many substances are neutral or non-ionic and their presence would not be detected by this means.
The metabolism of fish is non-mammalian and in many instances large variations in tolerance exist between fish and mammals.The fish monitors are subject to a high level of false alarms and this may lead to a disregard of a true pollution incident.
Discrete sample analysis is in general too slow in response for transient pollution incidents which may only last for a few hours.Discrete analysis may take up to thirty hours between collection and reportage of results.
The inconsistency in time scale associated with the various methods of reportage means that this method is usually unreliable for routine use.

Therefore,it is not inconceivable that transient pollutants could pass through the water treatment system and into the drinking water supply without detection by existing systems.

THE RISK TO THE POPULATION

In 1984 a spillage of liquefied phenol occured in the upper reaches of the River Dee in North Wales by an accidental spillage.The pollutant slug passed through the water treatment works without detection and was exported to a neighboroughing Water Authority where it was distributed to over two million consumers.The short term effects included nausea,malaise and recurrent headaches.Long term effects of the incident on the health of those affected have not yet been investigated.The pollutant remained in the drinking water system for over two weeks.

In 1988 several tonnes of liquid aluminium sulphate were accidentally placed in the treated water holding tank at the Lowermoor Treatment Works in North Devon.The water supply of twenty thousand consumers was affected.Seven thousand people consumed the polluted water and of these 60% sought medical advice for symptoms attributable to the pollutant.Several of those affected still suffer from the effects.A long term epidemilogical study has been promised.

In 1989 a sewage spillage north of Banbury in Oxfordshire passed through the treatment works undetected and affected the drinking water supply of several million people.Water had to be boiled prior to consumption.This situation lasted for several months in some of the affected areas.

The examples above have occurred during the past few years and represent the magnitude of consequence that a pollution incident may produce.

It is possible that some of the people affected by a pollutant will be affected for the remainder of their life.There is some evidence that long term effects could result from Camelford and some of those affected by cryptosporidium will remain so for the rest of their life or at least until an antibiotic is produced which can be used for it's treatment.

Susceptible Groups.

The likelihood that the injuries sustained in a rail or air crash will be independent of initial health state,age or vulnerability of the casualty is high.The opposite is likely to be true of the degree of damage to the human organism sustained in a water pollution incident.

Ingestion of an infective agent or toxic substance will produce varying degrees of severity dependant upon initial health,age and vulnerability.
The chronic sick,the metabolically disabled and the pregnant female are all very susceptible to these types of pollutants.The neonate and perinatal age group,if bottle fed,and the elderly are also susceptible.

People if between ten to fifty years of age,and if in average health,are more resilient to the ingestion of infective or toxic agents at low concentrations than are those groups mentioned above. However,everyone will succumb if the level of infective or toxic agent consumed is sufficiently high.

It is not unlikely that the consumption of sufficient toxic or infective agent would have an adverse effect on longevity,and,if a large number of individuals were so affected could this not then be termed a "disaster".

With many chemical substances the magnitude of the effects may not become apparent until several years after the incident.Chemicals with mutagenic tendencies may leave effects which will be manifest many years after the incident in the form of neoplasms.Certain substances may affect the developing ova,whether or not in a fertilised state,these effects will not become apparent until after birth.

THE RISK TO THE DRINKING WATER SUPPLY.

Not all pollution incidents fail to be detected and in practice the Water Authorities appear to be extremely efficient at containing the situation.But,there are instances where systems break down and the pollutants are not be detected until they have entered the drinking water supply.

The total number of water treatment works in England and Wales exceeds 1000.These produce 17,000 Ml of potable water per day(4),60-70% of that water is abstracted from surface waters(4).These surface waters are vulnerable to pollution from industry,agriculture and waste water treatment plants.

As stated previously,each of the main treatment works in England and Wales are liable to face,on average,5 - 6 serious pollution incidents every year,i.e.one every two months.This figure is quite close to the number of intake closures per year for large treatment works.Analysis of the Annual Reports of the Water Authorities indicate that the abstraction ports of major works are closed about 4 times per year on average to prevent pollutant slugs being drawn into the treatment process.The average large water treatment works supplies 125,000 people with drinking water.The average serious pollution incident rate for England and Wales is 6 incidents per day.Therefore,every day 750,000 people are at risk from the effects of pollution incidents.The risk to the individual in England and Wales is of the order of 1.5 x 10E-2.

METHODS OF PREVENTION.

The methods currently employed by the Water Authorities to treat raw water will not eliminate pollutant slugs.The current methods are targeted at reduction of bacterial count,oxygen content,acidity,colour,odour and taste.The methods currently employed are effective at the microgram per litre level whereas pollutant slugs usually present with concentrations of milligrams per litre.Therefore the pollutant passes through the system virtually unaltered in concentration.

When faced by a major pollution incident the only remedy available to the Water Authorities is to close the abstraction point until the pollutant slug has passed and this is the current practice employed.

After the River Dee incident the University of Bradford and the United Kingdom Atomic Energy Authority were requested to devise a means of assessing the risk faced by abstraction from the River Dee.From this study arose the concept of the Potential Abstraction Risk Index of Keller and Lamb(the PARI)(5).The PARI is based on the ratio of the potential concentration achievable in the river if a complete stock of a toxic substance should enter the water course within a period of twenty four hours.The PARI is the ratio of the potential concentration and the maximum acceptable concentration for man which is based on the toxicity of the substance and assuming that the average 70 kg person consumes 2 litres of the contaminated water in one day.The method has been further improved by Keller and Wilson(6).

The values obtained by the Acceptable Concentration calculation are different to those of the European Economic Commission(6) in that the latter are proposed for a life-times exposure whereas the Acceptable Concentration defined as above should not be exceeded for more than 24 hours.There is a safety factor within the Acceptable Concentration of 10,000 to allow for a high level of protection of the populace.

Each site within a river catchment area is surveyed as to the chemical stock holding maintained on that site.From this the PARI is calculated and the value ascribed to the site.If the site PARI is equal to and or below unity then the site is regarded as non-hazardous.Any site with a PARI greater than unity is regarded as a potential threat to the drinking water supply.

It is suggested that the following Hazard Ratings be used for classifying sites,

PARI	1 - 10	Low risk potential
PARI	11 - 100	High risk potential
PARI	> 100	Potentially catastrophic

The above ratings allow the sites to be ranked in a manner which is easily understood and is meaningful with respect to type and/or quantity of stock held.

Protection and Exclusion Zones.
The Control of Pollution Act (1976) gives the Water Authorities the power to create zones within river catchment areas within which they have powers to prevent and or restrict the storage or manufacture of potentially hazardous substances.These powers have been incorporated within the Water Act (1988) now laid before Parliament.

The PARI is an ideal technique for the creation of such protection zones in that it addresses both the toxicity and the stock holding level of the hazardous material.

During planning applications the PARI technique can be applied to restrict or prevent the stock holding of potentially hazardous substances within the catchment area of particularly vulnerable river courses.

Figure 3 shows the implication of the application of such zones of protection and exclusion for the River Derwent around Derby.

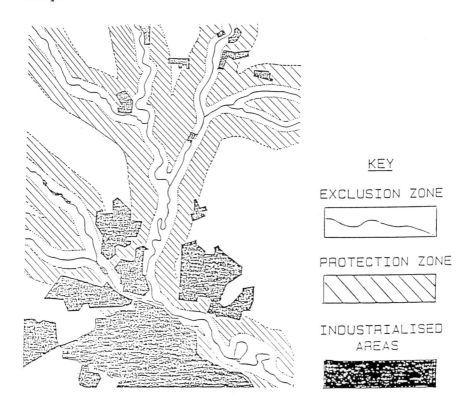

KEY

EXCLUSION ZONE

PROTECTION ZONE

INDUSTRIALISED
AREAS

MAP OF RIVER DERWENT SHOWING
EXCLUSION AND PROTECTION ZONES

FIGURE 3

The PARI method is being developed further to extend it's usefulness to classify industrial emissions, farm slurry and sewage spillages.

On the premise that prevention is better than cure it is recommended that a system such as the Potential Abstraction Risk Index be assessed as a working method for site classification and control.

CONCLUSIONS

From the high number of pollution incidents that occur in the rivers in the United Kingdom and the large number of water treatment works that exist it can be seen that the potential for pollutants to be missed and thus entering the water supply is high.

The current methods of water treatment were not designed to cope with the high concentrations of substances that occur in a pollution incident. Therefore, the pollutant passes through the treatment works virtually unaltered.

The methods of detection of pollution incidents can prove to be unreliable in certain circumstances and as early warning is important to Water Authorities the methods employed should have a higher degree of reliability.

The populations that are worst affected by pollutant contamination of the water supply are those who are most reliant on a supply being wholesome, i.e.the metabolically disabled, neonates and pregnant females.

RECOMMENDATIONS

1). That the Water Authorities and other responsible bodies examine classification systems for hazardous sites with the view to improving the reliability of hazardous material handling.

2). That reliable methods of early warning to Water Authorities in the event of a pollution incident be sought and implemented.

3). That possible effects from recent pollution incidents on susceptible populations be studied .

REFERENCES

1).Keller,A.Z.,1989,"Disaster Prevention Conference",Univesity of Bradford.Sept.1989

2).Wilson,H.C.,1989,"Disaster Prevention and Limitation Unit",Technical Report DPLU/4,University of Bradford

3).Chem Facts,United Kingdom (1987),Chemical Intelligence Services

4).Waterfacts 1988,Water Authorities Association

5).Lamb,B.,Keller,A.Z.,(1987),"The Identification and Quantification of Risks to the Public Water Supplies" in "Reliability 87",National Conference,Birmingham,1987

6).Keller,A.Z.,Wilson,H.C.,(1987),Environmental Pollution,1, Institute of Chemical Engineers.

DISASTER PREVENTION:
A CASE FOR THE FUTURE

R. W. Suddards, CBE,
Pro-Chancellor and Chairman of the Council,
University of Bradford

My first connection with Civil Emergencies arose with the Bradford disaster. The Chief Executive divided the tasks into three - the caring task, the raising of funds (which was brilliantly co-ordinated by Nigel Grizzard) and the administration of the trust fund, with which I was responsible.

My interest in Civil Emergencies was sparked off by a letter from Ms Sondra May Henderson in The Times when she told of her experiences in the Dover/Zeebrugge disaster. I felt that if half of what she said was correct (and I had no reason to disbelieve that all of it was not correct) it was a tragic state of affairs.

My response was to suggest a simple mechanism: in effect a sophisticated telephone directory so that all the experts in a particular subject were readily accessible in case of a disaster. It was important that alternative numbers be made available and there is obviously some judgmental basis to be considered as to who are the appropriate experts. Having said that, I still think it is a relatively easy matter to pursue. I was pleased that Desmond Fennell QC adopted this idea in his Kings Cross report.

I wrote to The Times with this idea and the result was a deluge of letters which led to the first conference at Bradford University in late 1987. Many of the people attending this conference were at the earlier conference. It was intended to be exploratory, but we came to some general conclusions:

(1)   A Civil Emergency is a situation which is likely to or which does result in a number of deaths or injuries and which is seen to need the services of more than one body of uniformed or voluntary service.

(2)   There is a strong need for a mechanism to ensure that all areas must have proper plans for Civil Emergencies, must keep them up to date, and must indicate to the public that such plans are in place even though the members of the public may not necessarily be entitled to inspect the plans.

Of course we recognise that many of the organisations which would be involved in a Civil Emergency have already in place plans for dealing with emergency. Many plans are sophisticated and take into account up to date knowledge. There may be a considerable variation in quality; there is no information that all bodies have such plans; there is no statutory duty to plan imposed on any authority.

But the fact that "something needed to be done" was sufficient to get us to take action. I saw the Prime Minister, and from this the Home Office was allocated the responsibility of looking at the whole problem of Civil Emergencies. This was fortuitous, because the Home Office is considered by some to be a graveyard for lost causes, but it equally houses enthusiasts in subjects.

We got an enthusiast, Bob Whalley, and subsequently his colleague, Richard Korniski. This was splendid, because both developed the thinking on Civil Emergencies enormously.

In all systems of control in this country preference is rightly given to operation by consensus. If society can be satisfied that a system of self-regulation is available and is likely to be enforced, then that is preferable to imposing more regulation by statute with powers through the criminal courts or by special enforcement procedures.

The facts that there are so many good plans and that we have such dedicated uniformed forces and voluntary services is very persuasive to a conclusion that we should continue to rely on the present unco-ordinated system.

The rest of the story is well known to us all. A Home Office discussion document was produced which had a wide circulation and which resulted in a lot of very positive suggestions being made.

My suggestions were based on the following propositions. I believe that the case for a compulsory code of co-ordination of the many disparate bodies which could be involved in planning the carrying out of emergency procedures is very strong and might be considered over-whelming. Such a code, backed by legislation, would further encourage the sensible, would discipline the backsliders and comfort the public. A mechanism of regular review of plans would ensure that developing knowledge and experience would be thoroughly understood and disseminated throughout the country.

It seemed to me that a duty to prepare plans and to co-ordinate action in an emergency should be laid primarily on the local authorities who would be required to bring together all the relevant uniformed and voluntary organisations who would by statute be obliged to be consulted. This would involve a statutory recognition of certain voluntary services which may be novel but would not, I imagine, be unwelcome to them.

Such activity must be co-ordinated nationally by some body on which would be placed a statutory duty of maintaining consistency and continuity in the framing and carrying out the local plans. Such a duty could be placed on Central Government, the Health and Safety Executive, one of the existing HM Inspectorates or on a new body specially set up for the purpose. I believe that it is best placed on an existing Central Government body and my preference would be for the Home Office.

When the Government did not accept the immediate argument for statutory control I was neither surprised nor dismayed. Statutory controls inevitably take a good deal of time. It was right that the two or three year period to achieving statutory control should not be wasted. It was encouraging therefore to have the indication that the Home Office would appoint a Civil Emergency Adviser.

I believe a major step has been taken by the Government in agreeing to this appointment. There is a huge amount of information available the world over, and this has to be assessed, analysed and swung into action at the appropriate time. The librarians of Bradford University have done a survey on the information that is available, and it is not only considerable, but in great depth.

This gives even more point to this conference and the unit being properly called a Disaster Prevention and Limitation Unit. We have all been involved in some way or another in dealing with disaster organisation. We must look to the stage behind, namely to prevention. Experience of disaster control is good and will help in disaster prevention, but we must emphasise the activity of control.

For these reasons I am happy that we have not gone head-long into legislation, although I am convinced that such legislation will be necessary eventually.

My only concern is resource. Lord Justice Taylor was able to produce his report on Hillsborough in the short time that he did because he had the resources of 10 firms of lawyers, 10 QC's, 10 Junior Counsel, the Police Forces, and his own back up from the Home Office. If that kind of resource is going to be given to the Civil Emergency Adviser, that is splendid. If it is not, then I think the whole process will be terribly slowed down.

Experience can and does help. When we set up the Bradford Disaster Fund we had certain maxims:

(1)   That we should be organised as a discretionary
      trust and not as a charity.

(2) That we should distribute the fund unequally as hardships had been suffered unequally.

(3) That we should not at any time publicise to whom we had made distributions nor how much any individual or category of persons had received.

(4) That we should pay out capital sums rather than regular income payments.

(5) That we should pay out to the beneficiaries as quickly as possible.

(6) That we as trustees should not meet any of the beneficiaries - because, we concluded, that if we saw one of the beneficiaries then we should see them all.

(7) That the costs of the administration be kept to a minimum.

All this was novel at the time. It has now ceased to be novel as we have advised a number of trusts which have been set up at Hungerford, Dover, Kegworth, Hillsborough, Lockerbie and so on. We are indeed having a further exchange of ideas of all these trust fund managers in November.

Every one of these funds add some knowledge or experience which can be passed on. This is perhaps the pattern to work to.

The news mentioned by Richard Korniski today that "new powers are being given to fire and civil defence authorities to deal with all disasters. Home Secretary Douglas Hurd announced plans for an overall disaster scheme in big cities like London and Manchester. Up to now, the authorities have been responsible for fire services and civil defence only in time of war. But the recent series of disasters has persuaded the government that they should take on the role of emergency co-ordination centres in the six metropolitan areas - London, Manchester, Birmingham, Liverpool, Yorkshire and Tyneside." That is good news. In the same way that disaster funds set up bring a little more knowledge, a little more experience, and changes in procedure, similarly emergency co-ordination information adds more and more each time there is one experienced.

Far from being depressed, I am enthused by the fact that the Government is doing something. The brief of the Home Officer Adviser and the major cities can include disaster prevention and limitation. This is excellent news. It is a pleasure to be present at this conference, which I hope adds more knowledge to the whole subject.

APPLYING SAFETY AND RELIABILITY METHODOLOGY FOR HIGH RISK INSTALLATIONS TO OTHER AREAS OF LIFE

C. Kara-Zaitri, Department of Industrial Technology,
University of Bradford

1. ABSTRACT:

This paper discusses the application of risk, safety and reliability techniques to disaster prevention and limitation. Techniques such as Hazard Analysis, Preliminary Hazard Analysis, Failure Mode Effect and Criticality Analysis, Fault Tree analysis, Hazard and Operability Analysis, Cause and Consequence Analysis and Common Cause Analysis are described at length and their suitability for disaster prevention validated. Other characteristics of risk and reliability studies such as the importance of automatic and manual data collection and the advantages of warning devices are also discussed. The paper also addresses the benefits and needs of databases for emergency planning as well as the problems of fusing both soft and hard data together to arrive at an accepted level of system behaviour. Some available databases such as FINDS and CRISIS are reviewed. Finally, new developments, such as Expert Systems, already validated in the field of risk and safety studies are explored and extended to disaster management.

2. INTRODUCTION:

In the 1940's, a group of German scientists led by Wernher Von Braun were developing the V-1 missile. The ten missiles which were manufactured initially, failed on the launching pads. An intense investigation was carried out and forced these scientists to think in a structured manner about the possible causes of failure. The outcome of this investigation was no less than the development of the well practised mathematical method known today as Reliability Block Diagrams. Similar studies were carried out in the

United states, and a subsequent increase in component
reliability and a decrease of number of catastrophes was
noticed. However, it is thought that this increase of
reliability was simply a direct result of an emphasis on
better quality. That era was marked by the considerable
interest shown by management in sampling plans for
inspection and control charts for variables and
attributes.

In the 1950's, the field of safety emerged as the most
popular and practiced discipline of the decade particularly
after the birth of aerospace and nuclear industries. This
decade was marked by the start of the use of reliability
terms such as "failure rate" and "life expectancy". This
era was also marked by the first attempts to quantify human
reliability in terms of number of errors per task.

The 1960's saw the beginnings of the Intercontinental
Ballistic Missiles and the man-rated rocket development
such as the Mercury and Gemini programs. These projects
were possible because new reliability techniques and a
large variety of specialised applications became
available. The new techniques that emerged included
Failure Mode Effect and Criticality Analysis (FMECA) and
Fault Tree Analysis (FTA). In FMECA, the modes of failure
of each component as well as the corresponding effects were
recorded and subsequently analysed. In FTA, a highly
undesired event was identified and a list of all
contributory events leading to the occurrence of the
undesired event was compiled. Furthermore, manual data
collection began. The data recorded included actual
failure times, inspection and tests results. This led to
the emergence of new rules and regulations in safety such
as the system safety study as an independent activity in
all major projects. This particular rule was mandated by
the US Air force following the disastrous accidents at four
ICBM complexes in 1962. This decade also saw the beginnings
of Military Standards such as MIL-STD-882, "System Safety
Programs for Systems and Associated Subsystems and
Equipment" and MIL-STD-471, "Maintainability, Verification,
Demonstration, Evaluation".

The 1970's were marked by the project entitled "WASH 1400,
The Reactor Safety Study". This project involved an
extensive risk assessment exercise sponsored by the United
States Atomic Energy Commission. In this project, a large
number of potential nuclear accidents were studied at
length. Professor Rasmussen, leading the research team,
numerically ranked each accident and assessed their
consequence to the public. Similar studies based on Fault
and Event trees began in Europe. The application of such

powerful techniques moved from the Nuclear industry to chemical and other industries. This was a direct result of rising public clamour regarding industrial hazards.

The 1980's were marked by a large number of disasters ranging from nuclear and chemical industries to transport and sport industries. These recent accidents involving enormous damages, loss of capital and life continue to be felt and are too well documented to need repetition. The 1980's were also marked by the birth of the Personal Micro Computer (PC) which revolutionized disaster and emergency research. Main frame computers were available in the 1970's but because of their high cost, their use was limited to large industrial organisations. PC's are now widely used in disaster prevention and limitation. A large number of databases directly related to emergency planning have emerged and can be accessed rapidly and efficiently by PC's via various communication media. Research in risk and safety has also significantly increased and major and complex software for FTA and FMECA is now widely available on Micro Computers. The demand for safety is stronger than ever. This era is marked by the birth of new legislations, new safety rules. The field of risk, safety and reliability is now linked to Artificial Intelligence, fuzzy logic and complex decision making.

3. APPLICATION OF RISK, RELIABILITY AND SAFETY ASSESSEMENT:

Every time a disaster occurs, the same questions arise; could the disaster have been prevented?, could the response have been faster? Could the handling of victims have been quicker? These questions lead to the following five activities in disaster prevention and limitation. These activities are as follows:

- Preventive actions: All actions that prevent the occurrence of a catastrophe.

- Hazard assessment and warning: These normally include monitoring and warning devices .

- Short term protective measures such as fire fighting, mobilisation of resources and people and evacuation.

- Long term protective measures such as design for radiation resistant structures.

- Risk assessment and public information

All these activities can be dealt with using some standard risk, reliability and safety techniques. It is not the purpose of this paper to discuss all the available techniques. Only a selection of methods useful for disaster management and limitation are presented. These are as follows:

1. HAZARD ANALYSIS (HA):

    This method represents a systematic and formalized way by which hazards can be examined. A hazard can be defined as a potential for causing injury, morbidity and damage to equipment or property. The method initially searches for potential candidate events or activities which might cause a hazard. Once these major hazards have been identified, corrective actions are then taken to control them to an accepted level.

    Hazards possess two properties namely severity and likelihood. These will be described in detail in Failure Mode Effect and Criticality Analysis. They fundamentally represent the degree of potential harm and associated probability value. Since quantitative data is not always available or appropriate, qualitative classes or levels are used instead. A third property that hazards may possess is the combination of severity and likelihood. This is termed hazard index and is generally calculated as the product of severity and likelihood classes.

    In the literature [1], suggestions related to a list of guide words which are used to stimulate the exercise of creative thinking. These words include "less of, more of, as well as, reverse". Hazard analysis is a useful tool for identifying potential hazards as well as primary events directly responsible for the hazard generation process. Once these primary events are established, their logical structure is modified in an attempt to reduce the hazard. Hazard analysis is basically a clear definition of the system itself and its boundaries in terms of potential hazards.

2. PRELIMINARY HAZARD ANALYSIS (PHA):

    PHA is the extension of the initial phase of hazard analysis. It is a more formal technique that examines all the previously identified hazards in great detail as well as corrective measures and consequences of the accident. One use of PHA is to determine which undesired events requires an even further investigation technique such as Fault Tree

Analysis. PHA work should start during concept and design stages of new systems. If the system is already in operation, PHA plays a major role in the safety evaluation program. PHA activities can be divided in three classes:

- Safety devices: These represent additional components, modules or systems for the purpose of safety.

  Warning devices: These are additional items required when safety devices cannot be utilised.

  Procedures and training: If both safety and warning devices do not work, operation procedures and training of operators is required.

PHA is more often than not based on a narrative basis and relies heavily on the importance put by the analyst on the horizontal inter-relationship of hazards at an item level but also on the vertical inter-relationships of modules and systems.

3. FAILURE MODE & EFFECT ANALYSIS (FMEA):

This method [2] represents an inductive analysis that systematically details on a component basis, all possible modes of failure and assesses resulting effects on the system. This technique is very well described by MIL-STD-1629A [3] which gives clear definitions of every word used in the method. In practice, reliability engineers record all failure modes and effects in the form of a table. The headings in the table vary from one industry to another. However, the following headings are found in the majority of FMEA studies [4]:

- Identification: This contains the unique identification of the item under consideration. The item can be a component, module or system.

- Function: This contains a clear statement about the intended purpose of the item in question.

- Failure mode and failure cause: This includes the description of the failure modes and causes. For each item, there may be more than one mode of failure.

- Failure mode frequency: This provides a quantitative measure on the frequency of the failure mode. This measure is normally expressed in number of failures per thousand hours. In a preliminary analysis, a quantitative measure is not always available and hence a qualitative estimate may be used. This can be high, low, frequent and infrequent.

- Failure mode effects: This contains information related to the consequence of the failure on the item itself as well as on the sub-assembly that contains that item. Here again, a tree structure is in fact used to look at each indenture level separately.

- Detection: This contains ways in which such failure is detected.

- Corrective measure: This contains a statement as to what corrective measure is necessary to restore the item to each normal working state.

- Severity: This represents the severity of the failure in terms of a category classification system suggested by MIL-STD-1629A. This classification is as follows:

  1. Catastrophic : Death
                    Loss of system
  2. Critical   : Severe injury or morbidity
                    Major damage to system
  3. marginal   : Minor injury or morbidity
                    Minor damage to system
  4. Negligible   : No injury or morbidity
                    No damage to system

The failure rate and the criticality of each component, module and system is evaluated on a hierarchical basis. The ranking of such failure modes in terms of importance helps the analyst identify potential events which represent suitable candidate top events for Fault Tree Analysis.

4. CRITICALITY ANALYSIS (CA):

This method [2] represents the second stage of FMEA and is referred to as Failure Mode Effect and Criticality Analysis (FMECA). The new information required is the likelihood of each event in terms of a probabilistic value when available or a probabilistic level otherwise. Here again

MIL-STD-1629A [3] suggests the following five levels of likelihood.

- Level A: Frequent. A high probability of occurrence during the item operating time

- Level B: Reasonably probable. A moderate probability of occurrence during the item operating time

- Level C: Occasional. An occasional probability of occurrence during the item operating time

- Level D: Remote. An unlikely probability of occurrence during the item operating time

- Level E: Extremely unlikely. A failure whose probability of occurrence is essentially zero.

Thus it is possible to specify both severity classes and likelihood levels for each item (qualitative or quantitative). The combinations of severity and likelihood can be conveniently taken into account in "the criticality matrix". The latter is a visual aid to identify critical items and hence areas which require further design, redundancy or reliability improvements. In addition, a criticality index can be computed for each item. The criticality index is rated in more than one way and can serve more than one purpose. For example, it could include human, operational and environmental factors.

5. FAULT TREE ANALYSIS (FTA):

The consequences of any disaster can be divided into three categories namely:

1. Human losses:
    - Death
    - Injury
    - Morbidity or disability
. 2. Economic losses:
    - Loss of capital equipment
    - Production shut-down
3. Environmental losses:
    - Pollution

These losses can occur as a result of basic failures. There are basically four types of failure:
1. Human failure:
    - Operator error
    - Design error
    - Maintenance error

2. Hardware failure:
   - Incorrect operation
   - Leakage of toxic fluid
3. Software failure:
   - Incorrect programming
   - Incomplete programming
4. Environment related failure:
   - Ignition caused by lightning
   - Storm
   - Snow

"Man made" disasters are frequently caused by a combination of these failures. The combination and causal relationship of such failures can be developed by a fault tree. FTA was conceived (1961) by H. A. Watson of the Bell Laboratories to evaluate the safety of an ICBM launch control system. The technique was extended and refined by the Boeing Company so that quantitative analyses can be performed. In the 1970's, significant advances were made in this area of work and a large number of FTA computer programs emerged. The technique is basically a Boolean logic diagram of the consequences of basic failures (primary failures) on the considered system failure (top event). It is a detailed deductive analysis that requires considerable system information. Deductive analysis reasons from the general to the specific.

The fault tree analysis method [5,6,7] is a systematic, graphical and descriptive tool that can be applied to disaster prevention because it can determine the sets of events capable of triggering the top event (undesired event). The method is most useful during the early design phases of new systems as well as operational systems.

The synthesis of FTA is the repeated question by the analyst about the cause of an event. The method has the facility of including all types of causes. The fault tree is structured in such a way that the sequences of events that lead to the undesired event are shown below the top event and are logically related to the top event by OR and AND gates. This process is carried through until basic events (which cannot be developed further) are reached. It is evident that the analyst must have a thorough understanding of the system in question. For example, the analyst or perhaps the team of analysts must be well versed in physics, electronics and human behaviour. This detailed and intimate knowledge is the result of long sessions with designers and operators in order to make the understanding of the design and operation philosophies complete.

Some of the values of FTA in disaster prevention are as follows:

- It provides a graphical aid giving those who are not directly involved with design changes clear visibility of system behaviour .

- It makes the analyst concentrate on Accident Related Events. Because of the deductive approach, all contributory events to the top event are also thoroughly studied.

- It provides both qualitative and quantitative analyses. This is particularly useful because it allows the analyst to prune only lowest-order basic event data and concentrate on only most important and probable events.

- It provides the analyst with insight into system behaviour. Because the method is so detailed, it forces the analyst to understand the system from various angles and much more beyond the level of the designer.

6. HAZARDS AND OPERABILITY STUDIES (HAZOP):

This represents basically an extended version of a FMECA study in the sense that it also includes operability factors. Each time a problem is identified by the hazard study team, it is immediately referred to the appropriate design team or support group. This process is restarted until all hazards are thought to be protected to an "acceptable level". By carrying out a good Preliminary Hazard Analysis, all major hazards have already been reduced or eliminated. New factors such as maintenance, operability, shut-down and startup can be examined and rectified.

7. CAUSE-CONSEQUENCE ANALYSIS (CCA):

This technique [8] was invented by RISO laboratories in Denmark. The method starts by identifying a critical event and then traces back all the resulting chain of events throughout the system. Since all conditions are taken into account, each considered path may take alternative sub-paths forming ultimately a subtree. The combination of all these subtrees in one single diagram represents the cause-consequence diagram. CCA is the fusion of Fault Tree Analysis (to show causes) and Event Tree Analysis (to show consequences). This technique is very complex and therefore the analyst must have a sound knowledge of Boolean Algebra, set theory and probability theory.

8. COMMON CAUSE FAILURES (CCF) OR COMMON-MODE FAILURES (CMF):

This method [9] is related to the study of failure of a system due to a common cause i.e. the loss, during a critical period of time, of functions or parts of functions due to an underlying common mechanism, fault or phenomena. Classic examples found in the literature include failure of aircraft jet engines due to bird ingestion. This method is normally practiced after FTA when all minimal cut sets (possible combinations of primary events leading to the occurrence of the top event) have been found. Common cause analysis [10] is then performed on each cut set to determine all the special conditions that closely link the considered primary events. Examples of special conditions are proximity, same manufacturer or same maintenance people. In CCA, secondary conditions are also examined. Examples of such conditions include vibration, pressure, temperature and radiation. The method combines special and secondary conditions under the same exposure area to arrive at a selection of potential common cause candidates. Each candidate is subsequently analysed and redesigned.

4. AUTOMATIC DATA COLLECTION AND MONITORING:

Manual data collection is an extremely important task but can be monotonous and tedious. In every day reliability and risk assessment studies, data sheets are designed and filled in. In practice, all the data recorded is coded into a form that is meaningful to a computer for subsequent analysis. With the advent of cheap hardware and software, this tedious task is no longer required. Data can be automatically recorded and subsequently analysed for various applications in risk and reliability studies. Analogue and digital sensors have revolutionized this process. First, a state of interest is identified whether it be temperature, voltage or vibration. This state is then examined and compared to either a high or low state i.e. logical output 1 or 0. Once the level is established, the output is then presented to the microprocessor for data storage. Data is collected over a certain period of time and analysed. Data analysis results are then tailored to suit the application of interest. At Bradford University, an automatic data monitoring system has been developed and extensively used in industry. The application of the latter system ranged from calculating the probability of failure of infant incubators to the computation of utilisation of high capital cost equipment in the NHS [11]. Some of the features of the system are as follows:

NIRMS [11]: (Non-Intrusive Reliability Monitoring Systems)

NIRMS is a system that has been initially developed for reliability and utilisation monitoring. An important feature of the system is the non-intrusive nature of the sensors. This aspect of the system is incorporated in the design in order to ensure that the equipment being monitored is not affected in any way by the monitoring system.

The system can be divided into two parts:

- The data logger: This comprises of an eight bit microprocessor unit (MCU) running in conjunction with a Real Time Clock (RTC), Read Only Memory (ROM) and Read Access memory (RAM). The design is CMOS design allowing a small battery to be utilised as a Power Supply Unit (PSU). The software is stored in read only memory (in this case, EPROM).

- The non-intrusive sensors: Their main feature is that they do not interrupt the operation of equipment from which data is being collected. Two types of sensors have been already been developed at Bradford University and they are the following:

  * Current sensor: This is capable of detecting current in three phase or single phase, armoured or unarmoured cable. No direct connection to the monitored equipment is necessary as the sensor detects magnetic fields.

  * Voltage sensor: It detects a voltage on an insulated conductor. Here again, no direct connection is required.

These automatic monitoring systems have proved to be a most useful tool for collecting accurately, effectively and most of all cheaply the required data in safety and reliability applications. It is thought that this expertise can be transported to disaster prevention where a certain characteristic or group of characteristics can be automatically monitored. A major use of such technique would perhaps be the constant monitoring of meteorological and radiation changes in order to prevent major disasters. All data collected should be sent to a central database for subsequent analysis.

5. WARNING DEVICES:

An aware and alert community can participate in the prevention and limitation of industrial disasters. Although such industrial crises are rare events, it surprising how many tell-tale signs and warnings that operators in particular and the community in general can use to initiate preventive actions. It is evident that the ability to recognize the importance of such warnings depends heavily on the awareness of the community on crises issues. These tell-tale signs can be thoroughly examined and, where appropriate, electronic devices can be developed to act as warning device in the case of a potential accident. The development of such devices requires a large volume of effort and a sustained dialogue between operators and emergency planners as well as designers and reliability engineers. Warning devices are widely used in industry. Perhaps the most common warning device is the fire detector. A number of studies [12] have been carried out by BURL (Bradford University Research Ltd.) on automatic fire detection systems as part of a risk and reliability exercise. Various factors such as evacuation times and false alarms were studied at length. The data recorded was collected from four different hospitals. The new generation of smart detectors (Autronica) was included in the analysis. BURL is also involved in the design of a CAtastrophy Predictor METer (CAPMET) system where data is recorded and modelled on the Extreme value distribution. This statistical distribution is based on worst case scenarios where only maximum values of a certain characteristic are taken into account. CAPMET is basically a system by which an undesired event can be predicted and its time of occurrence estimated.

6. THE NEED FOR DATABASES:

Research in disaster prevention and limitation requires the existence not only of hardware data but also the anticipated frequency of deviations from normal design environments and data relating to consequences of considered initiating events. Such databases should also include safety and warning devices as well as procedures and training of personnel in order to limit the effect of disasters. Consequence data is very much needed because if

available, it should include data related structural
integrity, meteorological conditions, toxicity, population
densities and environmental impacts [13,14]. Data directly
needed for risk and reliability assessment have existed for
more than a decade. For example, these databases include
failure information about electronic and mechanical
components. For disaster management, data collection is
relatively difficult and includes a wide spectrum of
information. The data collected could either be "hard" or
"soft" data. Hard data is related to information obtained
by physical tests and is normally easily collected. In
practice, the data is mathematically manipulated in order
to arrive at a mathematical or statistical model which
closely describes the process in question. Soft data
however, originates from the judgement of experts who are
very familiar with a particular equipment or process. The
difficulty with such data is that it is not always
rigourously quantifyble. The skill of the disaster manager
is to be able to fuse soft and hard data together to arrive
at degree of belief of system behaviour. Such skill can be
provided by the use of "Bayesian modelling" which is now
widely used in risk and reliability assessment studies. An
alternative would be to use Fuzzy Logic as explained in
[15].

Since the beginning of the 1980's, various databases
directly related to emergency planning and disaster
management have emerged. It is not the intention of the
present paper to discuss all of them but merely to examine
a few of them and discuss the usefulness and urgent need of
such information systems.

1. FIRE INFORMATION NATIONAL DATA SERVICE (FINDS):

    FINDS is a 24 hour computer link up between fire
    brigades enabling each one of them immediate exchange
    of requests for equipment and information. The
    software implemented is the result of a three year
    research programme. The system is operational by
    BURL (Bradford University Research Ltd.) and managed
    by by CACFOA (Chief and Assistant Chief Fire Officers
    Association). The service provided can be divided
    into two parts namely electronic mail and information
    pages. Brigades access the system on a daily basis
    to keep up-to-date with newly available information.
    Another main feature of the system developed is the
    possibility of using a mailshot in the case where an
    urgent information is required to be sent to all
    brigades. Since its launch in April 1988, FINDS has
    now proved to be a major tool to assist in the
    management and conduct of major accidents. In
    addition, subscribers as far as Melbourne (Australia)
    have shown significant interest.

2. CASUALTY RECORDING INFORMATION SORTING AND IDENTIFICATION SYSTEM (CRISIS): _

After the Bradford fire in 1985 [16], it was made clear that the work of the Casualty bureau in general and the identification of victims in particular would have been much easier had there been a purpose written computer program. Since then, a joint research programme by the West Yorkshire Police, Leeds University; Department of forensic medicine and forensic odontology and the computer software firm ISIS has been set up to develop such a computer system. ISIS was largely responsible for other similar projects such as MICA (Major Incident Computer Application) and HOLMES (Home Office Large Major Enquiry System). Essentially, CRISIS is a computer system by which detailed information about missing people is absorbed and subsequently compared with that of dead bodies. The system developed accepts data as code numbers or in plain English. The information stored is very much detailed and includes basic data such as physical features and complex data such as dental records for each tooth. CRISIS has already been used in recent major incidents including the Zeebrugge disaster.

3. COMPUTER GEOGRAPHICS:

This project [17] involved a grant of $ 1.6 M national research initiative into Geographical Information Systems (GIS) in collaboration with the Northern Regional Research Laboratory (RRL). The computer system developed includes maps with relevant statistics such as number of households, number of businesses and number of people. In the case of a major disaster, the system would be a powerful tool for identifying the disaster area and locate and warn the population under risk. The system is also capable of estimating the number of people as well as the shortest route between two points. With its visual display of dispersion of people and resources, this system represents an invaluable tool for a speedy and efficient coordination and response by authorities and emergency services in a disaster.

4. RESOURCES DATABASE:

In 1985, the Home Office funded a pilot study to construct a computerised database of resources allocation and availability. This study involved three English counties including Cambridgeshire, Northamptonshire and Cumbria. The aim of this exercise was clearly set out by the Home Office and is as follows [18]:

"To prepare a computerised register of resources as a model for use by local governments in emergency planning and in major emergency planning".

Ten main resource types were included in the database. These resources are Manpower, Water, Foodstocks, Buildings, Materials, Transport, Plant and equipment, Fuel and energy, Medical and environmental health and communications. A questionnaire was subsequently implemented and sent to various governments authorities. During and after data collection, many lessons were learnt. However, the study reinforced the view that the availability of information related to local resources is a vital aid for management in a crisis situation. One of the main recommendations of the study was to extend this exercise to every authority.

Obviously, lessons have been learnt about these various databases. The problem of combining all these databases still remains. With 1992 in mind, what is really needed is the establishment, at a European level, of a database which includes all resources for fire fighting disasters and a structured strategy that goes beyond responding to emergencies and includes forecasting, prevention and early warning as well as rescue operations, aid and short and long term reconstruction. These European information shells will establish a permanent network of liaison officers who will be responsible for updating, using and communicating data rapidly and effectively. Such cooperation of all European communities will be based on simulation, training and information. This cooperation will also strengthen contacts between emergency planners who might be called upon suddenly to work together.

7. EXPERT SYSTEMS (ES):

Several decades ago, scientists recognized that computers were not limited to numeric calculations but could also manipulate and process symbols. The application and use of such symbolic process led to several attempts to mimic the human mind. This manipulation of knowledge and its applications to human decision making and process learning is termed Artificial Intelligence. One of the major and practical applications of Artificial Intelligence are Expert Systems. These are normally bases or shells which interact with the computer in an efficient manner. Each shell can be interrogated using a standard dialogue system much as one would approach a human expert for advice. Obviously, before this type of consultation dialogue can be initiated, the knowledge bases must contain all pertinent information in terms of facts and rules. This transfer of expertise is not always easy because it deals with human opinions which are rarely free of vagueness and uncertainty.

An expert system is designed to emulate the reasoning process of an expert by using a model provided by the expert. The model takes the form of a tree or a number of subtrees. At the top of each tree, a goal is formulated. The model is designed to establish the certainty of the goal. The branches of the tree consist of a set of questions which the system will ask only if it requires that information. All Expert Systems use forward and backward chaining. When a question is asked and an answer is given, the information is propagated throughout the reasoning network and all hypotheses are automatically updated. The system may several levels of nesting.

Expert systems are thought to be a major tool for both prevention and limitation of disasters. When a disaster occurs, emergency managers have to make decisions accurately, efficiently and most of all rapidly. In addition, as explained by Dynes and Quarantelli [19], the number of decisions made and the information to be processed increases rapidly. Expert systems when appropriately used can validate the judgement of disaster managers and advice on the correct courses of action. Expert systems can also be used to perform certain functions automatically such as warning and communication. Work has already begun on the development of a prototype Expert System to provide decision support for emergency response in the nuclear industry (Salame et al. [20]). The information required for the shell was taken from the Indian Point Nuclear Power Plant and the State of New York Radiological Emergency Preparedness Group in the USA. This prototype system has been validated by comparing answers from past drills with real emergencies.

Expert systems seem to hold promise for implementation as a tool to be used for preventing and limiting disasters. With the new and advanced technology and in particular the use of mobile cellular phones and the availability of rapid and accurate digital information, as suggested by Belardo and Wallace [..], the role of Expert Systems technology in providing support to disaster managers is even greater.

8. CONCLUSIONS:

Disaster managers have in the past concentrated their research on natural catastrophes. Industrial disasters which are characteristics of modern societies now represent a qualitatively and quantitatively different type of crises. These disasters are generally a combination of human, organizational and technological disasters. They challenge disaster researchers to broaden their views and concepts of design, manufacture and operation. The emphasis is on forgetting the quick fix solution and spending more time looking for and developing new alternatives. This is precisely the object of risk, reliability and safety studies where every cause of failure and its associated consequences is examined under great detail. Understanding these causes of failure will provide a valuable tool for the prevention of future disasters and will serve as a basis for settling questions of safety, protection and liability.

9. RECOMMENDATIONS AND FUTURE WORK:

1. Risk and reliability studies already validated for large industrial installations should now be extended to smaller organisations and to other areas of life where a potential hazard may exist.

2. Disaster managers and researchers and emergency planners should have a sound understanding of risk and safety techniques and should be able to adopt them to study ways of preventing and limiting disasters.

3. Fusion of soft and hard data based on Bayesian modelling should be used to keep the consequences of disasters at a minimum.

4. Risk, relaibility and safety techniques should be combined with other disciplines such as inventory management and material control in order to arm the emergency manager with stronger tools for better emergency plans.

5. The development of databases management systems for the exchange of information on the relief structures and resources available on a national basis as well as an international basis should be carried out as soon as possible. Existing databases should be extended and included in a new network of information bases.

6. New networks for communications between the available databases should be established. This network of

information should assist the emergency planner to mobilize and assign, 24 hours a day, land and air-borne fire fighting teams from various countries.

7. Regular training procedures and simulation exercises should take place. Results and lessons learnt from these tests should be included in available databases so that evacuation and response time are kept at a minimum.

8. Community awareness, education and participation in exhibitions and other public events which deal with civil protection must be encouraged.

10. REFERENCES:

1. Robinson B. W. "Risk assessment in the chemical industry" RSA 5/78, CEC. Joint Research Centre Ispra Italy. 1978.

2. Humphreys M. "Failure Mode Effect and Criticality Analysis" Lecture notes 18. National Centre of Systems Reliability. UKAEA.

3. Military Standard 1629A. "Procedures for performing FMECA". Department of Defence. Washigton DC.

4. Human C. L. "The graphical FMECA" Proceedings of Reliability, maintainability Symposium. pp 298-303. 1975.

5. Haasl D. F. Advanced concepts in Fault Tree Analysis". Systems Safety Symposium. Seatle 1965.

6. Roland H. E. "Documentation for computer programs for Fault Tree Analysis" NHTSA 1980.

7. Lambert H. E. "Fault Tree Analysis" in decision making in systems analysis" Lawrence Livermore Laboratory. UCRL 51829 1975.

8. Nielson D. S. "The cause - Consequence diagram method as a basis for quantitative accident analysis" RIOS-M-1374. Danish Atomic Energy Commission. 1971.

9. Watson I. "Review of common cause failure". National Centre of Systems Reliability. R27. July 1987.

10. Burdick G. R. "COMCAN. A computer code for common cause analysis" IEEE Trans. Reliab. R26(2). 1977.

11. Kara-Zaitri C. "Utilisation of high capital/cost equipment in the NHS". University of Bradford. UBIT 127. December 1988.

12. Kara-Zaitri C. "Automatic Fire Detection Systems performance in the NHS. Data collection and analysis." University of Bradford. UBIT 126. December 1988.

13. Military Standard 217D. "Reliability prediction of electronic components". Department of defence. Washington.

14. "Electronic reliability data. A guide to selected components". 1981. National Centre of Systems Reliability.

15. Keller A. Z. & Kara-Zaitri C., "Further applications of fuzzy logic to reliability assessment and safety analysis". Microelectron. Reliab. Vol 29. No.3. pp 399-404. 1989.

16. "Post Bradford disaster hot line proposal". Civil Protection. Spring 1988.

17. "Computer Geographics". Civil protection. Winter 1988.

18. Kidd S. et al. "Construction and utilisation of a resource database in disaster planning". Disaster management. Vol.1, No.1, 1988.

19. Dynes and Quarantelli. "Organisational communication and decision making in crises". DOD ARPAN 00014-75-C-0458. Advanced research project agency. Washington. 1976.

20. Wallace W. A. "On managing disasters". Technical report No.37-87-118. Troy New York. Rensselaer Polytechnic Institute.

THE HUMAN ELEMENT IN DISASTERS

A. R. Hale, Safety Science Group,
Delft University of Technology,
The Netherlands

INTRODUCTION

The 1980s have already been rich in disasters and in the last
year or so the United Kingdom has sadly been particularly hard
hit. The catalogue of names includes: Piper Alpha, Hillsborough,
Clapham, Lockerbie, London Underground, Herald of Free Enterprise, to name only a few. The rest of Europe has provided its
fair share; the Airbus at the French, and the Italian flying team
at the Ramstein airshows, Heizel, Chernobyl, Sandoz, the
regularly repeated forest fires in the South of France, etc.
Further afield both developing countries (e.g. Bhopal, repeated
ferryboat disasters in the Philippines) and developed countries
(e.g. Challenger) have contributed their fair share.

All those who study disasters are agreed that the human and
organisational elements in both preventing and coping with
disasters are the dominant ones. It is the complex interaction
of these with the technological factors which determine on which
side of the fine line between control and disaster any system
falls. In this paper I want to give two examples of this complex
interaction and to draw some conclusions about the ways in which
the human element needs to be considered and influenced in trying
to prevent future disasters.

Let me make it clear at the outset that I am concentrating on
disasters in man-made systems and not so much on natural
disasters. The distinction is that, in man-made systems, we have
potentially the power to control the hazards and to prevent them
leading to harm, while in natural disasters we do not. The
distinction is not absolute however; even with natural disasters
such as the floods in Bangladesh or in 1953 in the Netherlands we
have the power to predict the weather, the power to make the
flooding worse by deforestation or to control it by building and
maintaining the Delta works, storm flood barriers and dykes.
With man-made disasters we are sometimes surprised by failures we
did not know could occur (Thalidomide, metal fatigue in the early
Comet aircraft, etc) and so did not attempt to control. When we
consider any disaster we therefore have to consider all the steps
in a problem-solving model (Hale 1985, figure 1). It is almost
superfluous to point out that all of these steps must be carried
out by individuals or organisations and that they must be
coordinated if a successful outcome is to be reached. The

importance of the human element in disaster prevention should
therefore already be clear.

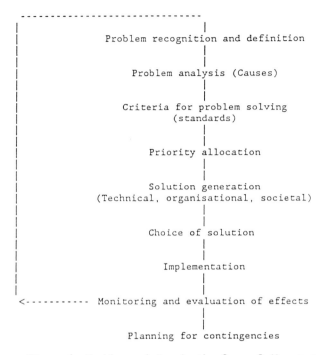

Figure 1. Problem solving in the face of disaster potential.

The greatest potential for man-made disasters exists in complex
systems such as high technology industry, road, rail and air
transport, and leisure activities where large crowds assemble;
here it is harder to develop and retain an oversight into all of
these problem solving tasks and their results. Matters are made
worse where complexity goes hand in hand with what Perrow (1984)
has called close coupling of the system; the property by which
events in one part of the system lead rapidly to consequences in
other, often unrelated parts of the system, such that small
deviations can suddenly lead to large and uncontrollable consequences.

ACCIDENTS IN COMPLEX SYSTEMS

Let us look at two examples of accidents which can be considered typical of safety problems in complex systems.

1. Herald of Free Enterprise

At 18.28 on the evening of March 6th 1987 the roll-on roll-off ferry Herald of Free Enterprise capsized in shallow water just outside the breakwater of Zeebrugge harbour with the loss of 188 lives. It had set off 23 minutes earlier with its bow doors open and trimmed slightly bow down. When the captain put on speed and turned as normal onto his course after leaving the harbour the ship scooped up water into the large open car deck, became fatally unstable and within 5 minutes had capsized.
The design of such ferries is inherently unstable. If water enters the open decks in sufficient quantities, either in the way it did here, or if the ship is holed in a collision (as had happened with other ferries on previous occasions), the ship rapidly turns over. The design is however regarded as essential for fast loading and unloading, which would be very difficult with the normal watertight bulkheads normally found in cargo ships.
The Herald had a chronic list to port which did not help matters. The bows had to be trimmed down to allow the ramp at Zeebrugge to reach the top car deck and the pumps to pump the water out of the ballast tanks were of a very small capacity, so that it took more than an hour to readjust the trim. This was always done while the ship was under way. The scuppers on the car deck were also inadequate to clear water streaming in. The technical and design factors therefore set the scene for the accident.

But, how did the bow doors come to be left open? The official enquiry found that the assistant bosun, whose job it was to close them had fallen asleep on his bunk after being busy with main-tenance and cleaning work. His boss, the bosun did not consider it his job to shut the doors or to check that they were shut. There was a lack of clarity between the chief officer and second officer over who was in charge of the loading operation. The second officer had been delegated to do it while the chief officer was elsewhere, but heard the chief officer giving instructions on the radio and assumed he was back and again responsible. The chief officer saw 'someone' coming from the direction of the controls of the doors, and assumed it was the assistant bosun (it was probably the bosun) and therefore assumed the doors were shut. He was then distracted by passengers. The chief officer then went to the bridge. The captain assumed, because there was no positive report to the contrary, that nothing was out of the ordinary. Company orders appeared only to require negative reporting (a procedure that fails to danger). The captain did not ask for a report and the chief officer gave

none. There was pressure to sail early, also confirmed in
company orders from the operations manager.

The captains of several of the ferries had requested an indicator
on the bridge to show the state of the bow doors (cost c.£450),
but this had not been installed. The state of the doors was not
visible from the bridge, only from outside on the flying bridge.
There was also no instrument to indicate the trim of the ship in
relation to the load line; but nobody ever bothered to check this
anyway.

A series of procedural and human errors at several levels
introduced the conditions in which the technical and design
failures could produce their fatal result. Above all this sat a
management, criticised by the inquiry, which seemed to pay more
attention to punctuality and cost than to safety.

It was fortunate that the capsize at Zeebrugge occurred in
shallow water so that the ship could not turn further onto its
head. This meant that there was a possibility to escape from
the enclosed decks. The capsize also occurred relatively close
inshore and in reasonable weather so that rescue services could
be quickly on the scene and could work without too many problems.
Remarkably little panic occurred in the aftermath, and both crew
and passengers showed both rational and sometimes creative
behaviour in getting people out of the ship.

2. Challenger

On 28th January 1986 the Challenger shuttle was launched at 11:38
from Kennedy Space Centre. One minute later it exploded killing
all seven crew members. The O-rings on one of the joints of
righthand solid fuel booster-rockets failed as the shuttle passed
through a very violent wind shear. This allowed a plume of flame
to escape and play on the skin of the liquid hydrogen tank, which
ruptured, precipitating the explosion.

The potential for the failure was detected 9 years earlier in
initial tests of the solid fuel boosters. An earlier design
showed unexpected behaviour in tests. The joints expanded
instead of contracting during firing, and the O-rings became
unseated. The designers from Thiokol, the producers of the
boosters, persuaded the NASA that this was not critical, and
despite the fact that another company designed a technical fix
for the problem for another booster, Thiokol did not incorporate
this. Due to poor communications within NASA and between NASA
and the various suppliers no action was taken on the information,
discovered during other test phases, that the joints could spring
open and allow flame to get out. Under the pressures to meet the
launch deadlines the failures which were noticed during tests
from 1981 to 1985 were rated as acceptable, despite the fact that

joint failure had been given in 1981 the highest criticality rating (failure could cause total loss of craft and crew). No connection was made in the minds of those responsible between many of the indications during test and the possibility of the failure. This connection was finally made by the NASA booster project manager in 1985 and he imposed a ban on launching if there was any worry about top criticality items, including the joint leaks. There was, however, the possibility of granting a waiver of this ban if those concerned were sure the problem would not arise under the prevailing flight conditions. This waiver was frequently granted for tests thereafter on these grounds and top management at NASA did not query this fact.

Meanwhile new boosters had been ordered based on the technical fix discovered 4 years earlier. Therefore the director of the project at Thiokol regarded the problem as solved and indicated as such on reports to NASA. However the old boosters were still being used for the Challenger.

Despite an attempt by a Thiokol engineer to warn of the disaster potential in July 1985 a meeting of the engineers did not single this out as a critical item. The issue was also not noted by NASA because the senior manager missed the meeting. All of these communications problems were aggravated by an autocratic management style in a part of NASA and the justified fears that the suppliers would not get their contracts renewed if they made too much of their problems. There was a consequent fear among those in the various organisations of 'rocking the boat'. Responsibilities for the success of the launch programme and for safety were also not clearly separated.

On the day before the launch the temperature was unusually cold, which made the O-rings particularly stiff and more likely to fail. At this late stage the Thiokol engineer tried to persuade NASA for these reasons to stop the launch, but the objections were not taken seriously by the lower echelons and were not passed to the senior officials. Great pressures within NASA pushed the launch ahead in order to score a success with the administration and to make sure that the shuttle (and its teacher in space mission) was in orbit to coincide with President Reagan's State of the Union message on 28 January.

The pressures of high politics, high stakes and tight deadlines had created the conditions where poor communication within and between big organisations could prevent a technical risk being evaluated and assessed properly. Hence it manifested itself fatally.

No escape for the crew was possible. In the redesigned shuttle this has been incorporated at great cost, so that the shuttle can be blown clear from the boosters and land safely even in such an emergency. The explosion occurred over sea and hence no danger to others arose; the launches are deliberately made so that this is the case.

Conclusions

These are only two accidents out of many, but they show the characteristics of accidents in complex systems. There is no single factor or error which can be pointed to as the cause. Many factors combine, each on its own comparatively innocent. The organisation, frequently through inadequate planning and communication, propagates and compounds the faults. After lying latent for a time these faults can propagate with alarming speed, particularly in tightly-coupled organisations where the technology being used is sophisticated, and the situation is intrinsically unstable. No single person has a clear overview of what is going on and of how all the threads of the disaster are coming together; or if they do, they are not in a powerful enough position in the organisation to make their viewpoint prevail. The final trigger is often remarkably trivial and sometimes a pure chance deviation from normal circumstances.

To prevent the disasters occurring again it is not sufficient to point the finger at one person and criticise him for having failed to do what he should have done. The traditional view of safety and responsibility enshrined in compensation legislation is hardly appropriate for such cases. While individuals can be considered guilty of lack of thought, lack of foresight and even of breaking rules, it is hard to say that they have been grossly negligent. It is necessary to take a much broader and deeper look at the whole organisation and raise its level of systematic concern for safety a quantum level (or in the case of NASA to restore it to its previous high level). The Rogers Commission into the Challenger disaster, and the Court of Inquiry into the ferry disaster both made many detailed recommendations aimed at this.

PATTERNS OF DISASTER

Each disaster that we study is unique, but it is becoming increasingly clear that they have common features. Their uniqueness hides a fundamental structure in which human and organisational factors play a crucial role. This structure can be summarised in a model (Hale & Glendon 1987, figure 2).

1. Planning and learning
The crucial stage in this model is the first; the choice and design of the system itself. At this stage the potential disasters have to be identified and plans made to eliminate them or keep them under control. As the feedback loop in the model suggests, we can and must learn from disasters which have already occurred and we must do that systematically. But that is not sufficient; the disasters which catch us out always have that element of the new in them. They have usually never occurred quite like that before and so it takes creative thinking to predict them.

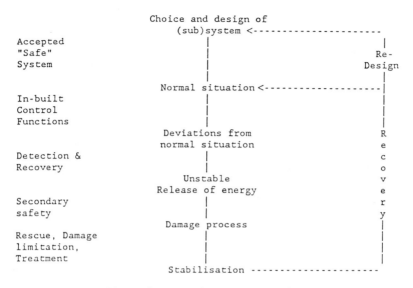

Figure 2. Deviations and Controls.

Creative and systematic thinking does not occur unless someone is
given the time to sit down and indulge in it. That someone also
needs sufficient knowledge of the system as it is designed and,
more importantly, as it will be used (and misused) in practice.
They also need the detachment to be able to think as broadly as
possible and not to be railroaded into rapid decisions about
modifications to be made to prevent the last disaster.

The immediate aftermath of the Hillsborough disaster saw crowd
control fences being removed by football clubs with as great a
speed as they had been put up after disasters such as Heizel.
Change and action are psychologically and politically necessary
after a disaster; some change is technically necessary to prevent
a recurrence; but change in systems is the greatest factor in
introducing new hazards which may have unpredictable and sometimes more serious consequences than the original evil they were
designed to combat (Johnson 1980). A human characteristic which
disaster prevention must control is therefore the tendency only
to look back after disasters and not to look creatively forward
to assess the negative as well as the positive aspects of
proposed changes. Another human factor is the tendency to be
strongly influenced by recent events and to forget those longer
in the past; hence precautions introduced after one disaster
gradually fall into disuse because there is no longer anybody in

an appropriate place in the organisation who experienced that problem. The corporate memory of an organisation can be far shorter than an individual memory (Kletz 1988). To overcome both of these problems we need better ways of systematically analysing and storing the lessons from old problems and the effectiveness of the solutions which we have tried in the past. This suggests that we need professionally trained watchdogs in an organisation and at local and national level to fulfil this function.

In the past we, as a society, used to adopt the viewpoint that people designing and making products and systems could be assumed to have thought about their dangers and taken appropriate action. We also assumed that the "customers" who bought the product or used the system could see for themselves whether this had been done and choose not to buy or take part if they were not satisfied. If something still went wrong, the designer or the employer could be punished or made to pay and that would encourage others in the future to do better. It is increasingly clear that this does not sufficiently control disasters. Even the additional policy of capturing all the available past experience in detailed rules and standards does not give sufficient guarantee that new elements in complex systems will not lead to disasters. Increasingly we are turning to the requirement to provide positive proof that a "designer" has thought sufficiently about potential dangers and has planned accordingly. Nuclear installations have been subject to this sort of licensing regime for many years. Major chemical plants in many European countries now fall under a similar requirement both in terms of prevention and in terms of their disaster plans (Post Seveso Directive). Medicines for many years, and more recently all chemicals which are brought onto the market in bulk have had to be shown, by extensive tests, to have an acceptably low risk of a wide range of effects. The recent European Directive on Machinery Safety requires manufacturers to set up a technical dossier including an analysis of all potential hazard scenarios (in installation, testing, maintenance and demolition as well as normal and predictable abnormal use). In the Netherlands all companies with more than 500 employees have to submit details for approval by the labour inspectorate of their safety services and the way in which they monitor and influence the company safety policy and practice. Similar, but more stringent, requirements relating to the total risk control system operate in Norway.

All of these provisions are designed to prove that the systematic and creative prediction and assessment of risks has taken place. A logical step would be to extend this to all systems which have such a disaster potential, whether they are industrial or not. The ingredients for disaster in major leisure activities (sport, leisure centres, fairgrounds, pop festivals, etc.) has been sufficiently demonstrated to indicate them as prime candidates for such assessment and licensing by professional and competent risk analysts.

2. Preventing deviations

The Herald of Free Enterprise case illustrates typical sorts of human behaviour which lead to deviations and ultimately to disasters. Such problems have been the subject of intensive research in the last decade (see e.g. Reason 1988, Reason 1989, Hale & Glendon 1987, Rasmussen et al. 1987, Hoyos & Zimolong 1989) and the results are too numerous to be summarised here. Two related problems can be singled out to illustrate how far we have come and how far we still have to go.

Many disasters are the result of a misdiagnosis of a problem. The Three Mile Island nuclear accident was the result of operators misinterpreting a leak in the cooling water system for an overpressure in the system and hence overriding the safety measures which would otherwise have brought the situation rapidly under control; Bhopal was partly due to misdiagnosis of the causes of heating in the storage tanks; Clapham involved an understandable (but wrong) decision to stop the train because a signal was giving a misleading message; at Hillsborough decisions were made to open the gates to relieve a threatening situation at one point without taking full account of the consequences of that action elsewhere.
People in many of these disasters chose, often for very good reasons to follow what turned out to be the wrong rule in arriving at their choice of behaviour.

Other disasters result from situations in which rules have fallen into disuse. In the Herald disaster the rules about checking the door closure were not followed; at Chernobyl the experts thought they could get away with doing tests under conditions which were expressly forbidden according to the rule book; a major accident in fog on the Dutch motorways occurred when many drivers ignored rules to slow down in poor visibility.
Usually the disaster does not occur the first time the rules are broken, but after that breach has become hallowed by much repetition and is accepted by the system as normal practice.

Both these problems highlight the way in which people behave in relation to rules; sometimes following them when they should not and at other times not following them when they should. The traditional response to disasters has always been to devise new rules which would have prevented that disaster and to impose them for the future. As a result old technologies usually have huge and unwieldy rule books. Safety rules are in essence about the specification, communication and control of safe behaviour in dangerous situations. In that respect we should all be devotees of safety rules. But the term also frequently conjures up very negative associations. Safety rule books have the image of unworkable lists which are devised by lawyers to pin the blame on someone after something goes wrong, rather than contributing to preventing things going wrong in the first place.

This latter view is typified by the results of a study by Elling (1987, 1988) among 50 railway workers which looked at their reactions to the rules governing work on and near railway lines.

- 80% considered that the rules were mainly concerned with pinning blame
- 79% thought there were too many rules (but 12 % thought there were too few)
- 77% found them conflicting
- 95% thought that work could not be finished on time if the rules were all followed
- 85% found it hard to find what they wanted in the rule book
- 70% found them too complex and hard to read
- 71% thought there was little motivation given to follow them
- not one could remember ever having referred to the rules in a practical work situation

Company safety rules in Elling's studies were seen as static and rigid, restricting and stifling initiative, and often not relevant to the situation facing the operators in practice. This emphasises the normative nature of company rules as things imposed from the outside by someone else, not always with the wholehearted consent of the governed. Imposed safety rules are often seen as in conflict with other imposed rules of a higher priority (e.g. production), or as applying only to some situations and not being possible to follow in the many exceptional situations which make up reality.

The last two decades have seen major attempts to deregulate safety at a national level; to replace rigidly worded laws with flexible frameworks of enabling legislation (e.g. the British Health & Safety at Work Act 1974 and the Dutch Working Environment Law 1980); to move from highly detailed standards to frameworks of essential safety objectives (e.g. the European Community Machinery Safety Directive and the draft CEN Machinery Safety Standard). These moves were taken partly to make legislation more understandable (GB 1972), partly to cope with the dynamic nature of technology and hazard prevention (Anon 1985). But we cannot do without rules at some level, otherwise there is no way to judge, before an accident happens, whether the organisation or individual is working safely.

This paradox of the positive and negative aspects of safety rules is one which we have not fully resolved (Hale 1989) and I see a major effort in research being necessary to establish when what sort of rules are positive and when negative. When do they stifle necessary creativity and freedom in acting safely and when do they prevent careless and thoughtless risk taking?

Meanwhile we need to act on our limited knowledge of how safety rules operate in systems. This tells us that such rules must be made by, or with the close participation of, those who will have to obey them. It also tells us that rule systems must be constantly reviewed and adapted to avoid them falling into disrepute and disuse. This is another task for the trained safety professional at organisation, local or national level.

3. Reacting to disaster

The third area where human behaviour plays a vital role in determining the extent of disasters is in reacting to the early stages of the disaster, the release of the harmful energy. This is the stage of evacuating buildings or sports stadiums in the face of fire, of villages or suburbs in the case of flood or radiation leak, the stage of sheltering from toxic releases and of rescuing survivors from collapsed buildings or the wreckage of transport accidents.

There is some research (e.g. Canter 1980) into behaviour in fires, which shows the behaviour to be largely predictable (i.e. seldom characterised by panic), but not always optimal for the saving of life. Thus people will risk themselves to rescue or be reunited with their family (as at Summerland) or their possessions; people will often not believe initial fire warnings until they have gone to check for themselves; they will use the familiar, rather than the closest exits from the building; they underestimate the dangers from smoke relative to those from heat; etc.

There is also some research into the responses to natural disasters (e.g. Lowenthal 1967), which suggests that people are reluctant to follow instructions to evacuate their homes in the face of warnings of flood and hurricane and often underestimate the threat, based on the fact that "it went OK last time".

Reports of the aftermath of disasters also show that the scene of the disaster, even before it is brought under control, is a powerful magnet for those not immediately affected. They come to offer help, but also to watch and stare and get in the way.

Such research is very patchy and often anecdotal upto now. We are still not at the stage where the behaviour and how to influence it can be systematically predicted. Above all the disaster plans which are increasingly being drawn up by companies and by local and national authorities are not assessed against the behavioural criteria which we can derive from theory and research. Will people really go indoors and calmly listen to the radio in response to sounding the warning sirens, as the instructions from the local authority where I live require? How many other people have actually found and read those instructions printed in the front of the local town guide? Will people barricade themselves in their cellars in response to a nuclear alert? When will they stay indoors and when will they attempt to flee in their cars after hearing that there is a major toxic chemical release?

All of these are vital questions if we are to minimise damage after the disaster has occurred.

CONCLUSION

This presentation has only skimmed the surface of the human and organisational element in disaster prevention and planning. Nothing has been said of risk communication and of training people to control or flee from disaster. Little has been said of the management of risk within organisations and of the co-ordination of disaster teams. Even the topics touched on have only been sketched in outline. I hope, however, that it is clear that no disaster prevention is possible without a systematic and integrated study of the human with the technical elements in systems. I hope also that the case has been clearly made for insisting that those who design and manage systems with a significant disaster potential should positively prove that they are capable of designing and managing them; that they have taken account of the predictable behaviour of people as both cause, potential preventor and most important victim of those disasters; and above all that they employ the necessary trained professionals to carry out periodical reviews and critiques of their plans and rules to avoid them becoming ossified and to keep them up to date with the lessons from research and from the disasters which still (sadly) occur.

BIBLIOGRAPHY

Anon. 1985 A council resolution over the new approach to standardisation and harmonisation. EEC Community Journal. 7 May.
Canter D.V. (Ed) 1980. Fires & human behaviour. Chichester. Wiley.
Elling M.G.M. 1987 Veilig werken volgens geschreven procedures: illusie en werkelijkheid. Communicatie in Bedrijf en Beroep. 2. 133-143
Elling M.G.M. 1988 Duidelijkheid, bruikbaarheid en effectiviteit van werk- en veiligheidsinstrukties. Tijdschrift voor Taal-beheersing 10 (1) 1-13.
Great Britain 1972 Safety and health at work: report of the Committee 1970-1972 vol 1. HMSO London.
Hale A.R. 1985. The human paradox in technology and safety. Inaugural lecture, Delft University of Technology.
Hale A.R. 1989. Safety Rules OK? Possibilities and limitations in behavioural safety strategies. International Conference: Strategies for Occupational Accident Prevention, Stockholm, 21-22 September 1989.

Hale A.R. & Glendon A.I. 1987. Individual behaviour in the
    control of danger. Amsterdam. Elsevier.
Hoyos C.G. & Zimolong B. 1988. Occupational safety and accident
    prevention: behavioural strategies and methods. Amsterdam.
    Elsevier.
Johnson W.G. 1980 MORT safety assurance systems. Marcel Dekker.
    New York
Kletz T. 1988. Learning from accidents in industry. London.
    Butterworth
Lowenthal D. (Ed). 1967. Environmental perception and behaviour.
    University of Chicago.
Perrow C. 1984. Normal accidents. New York: Basic Books.
Rasmussen J., Duncan K.D. & Leplat J. (Eds). 1987. New technology
    and human error. Chichester. Wiley.
Reason J.T. 1989 Human error. Cambridge University Press.
Reason. J.T. 1988. Summaries of 5 case studies illustrating the
    part played by latent failures in accident aetiology. Dept.
    of Psychology, University of Manchester.
Rogers W.P. et al. 1986. Report of the presidential commission on
    the space shuttle Challenger accident. Washington: NASA

MASS DISASTERS —
THE PATHOLOGISTS ROLE

Wing Commander I. Hill, OBE, RAF,
Department of Aviation and Forensic Pathology,
RAF Institute of Pathology and Tropical Medicine

INTRODUCTION

In recent years mass disasters appear to have received greater
media coverage than they did in the past and their frequency
seems to have increased. There has undoubtedly been a change
in the ways in which relatives and friends approach the
investigation of a mass disaster and its subsequent
proceedings. Arguably one of the biggest changes has been the
spectre of criminal prosecution, which has been raised,
particularly in the wake of Zeebrugge.

These changes have, as would be expected, had an effect upon
the pathologist's part in the investigation, however the
fundamental reasons for launching such investigations have not
changed and, apart from technological advances, the techniques
involved have remained the same. The basic objectives are to
establish the cause, or causes, discover the effects and
ascertain how these came about and to suggest ways in which
recurrences might be avoided. It has to be accepted, no matter
how unpalatable this may seem, that accidents will happen and
therefore, having established the facts, it is part of the
pathologist's task to suggest ways in which the effects of
future accidents may be ameliorated. These objectives can only
be achieved if all of the investigators work as members of a
co-ordinated team. A team in which there is mutual respect for
the respective skills of those involved. Recognition of this
fact is important for without it and a harmonious working
relationship the investigation may be compromised. Whilst this
paper looks at the pathologist's role, it should only be seen
as a part of the whole investigation and not as an entity
itself.

METHODS

There are essentially three ways of investigating accidents, a
central authority which investigates them all, a central
reference organisation which reviews the findings of others and
a mixture of the two. It could be argued that the first of
these methods is the ideal. Such a system allows a greater
body of experience to be built up, greater control can be
exercised over the investigation and therefore the information
yield ought to be greater. This system has been used in the
United Kingdom for the investigation of aircraft accidents for
many years. Central reference authorities are used in some
countries, especially in those which have a federal system of
law and organisation and in which long distances may have to be
travelled. The combined system is a useful compromise.

However the investigation is organised and whoever carries it out, it is important that a systematic approach is used. The pathologist must examine the site and the wreckage. The two may not always completely coincide with one another, thus in an aircraft accident there may be damage to trees, for example, long before the actual crash site. Signs such as these may give important clues as to the causation of the injuries as well as the accident.

Accident sites vary greatly. The wreckage trail in the Air India accident extended over five miles. In other cases the whole site may be confined within a few feet. No matter how large or how small the site is the same basic approach has to be applied if viable results are to be obtained. It should be divided into squares of about five metres by five, depending upon the size of the scene and the amount of debris present. If there is a large amount of wreckage in a particular area then it may be advisable to reduce the size of the search area. Ideally the pathologist, accompanied by a Scenes of Crime Officer and other specialists as necessary, should search each part. This may not be possible, especially if there are injured people in the wreckage, for they must be removed first. Everything must be photographed in great detail and the photographs must carry an identifying mark so that they can be placed in their proper context. When a body is found its position should be noted. Its relationship to the wreckage should be recorded and photographed. The body must be numbered. Pre-numbered labels must always be used. If they are not and there is more than one team examining the site, there is the ever attendant danger that there will be duplication of numbers, which will lead to considerable confusion and seriously undermine the value of the investigation. Two labels should always be attached to each body, preferably at some distance from one another. This gets around the danger of loss of a label during removal of the body. This is especially likely to happen when the conditions are hazardous. When there has been dismemberment the separate parts of bodies must always be labelled separately. There can be no guarantee that parts found separately belong to the body nearby. Similarly belongings, such as handbags, which are found close to a body should be removed separately and not placed with an adjacent body because such items may be thrown about by accident forces, coming to rest a long way away from their owners.

Examination of the accident site, which involves careful wreckage analysis, the positioning of bodies and survivors relative to the debris, is a lengthy process. It has to be done painstakingly so that nothing is missed. Every detail must be recorded, for once it has gone it has been irretrievably lost and so the overall value of the accumulated evidence is lost. This may mean that the disaster cannot be properly evaluated.

In the mortuary the bodies should be photographed again before their clothing is removed. Any special features should be noted and photographed separately. The bodies should be photographed again after the removal of clothing, again noting and photographing special features.

It is advisable for the pathologist to remove the clothing and personal belonging, which should then be handed to the police. In this respect all that is being followed is the routine practice of the examination of bodies found in suspicious circumstances. The practice of removing clothing and searching the pockets before the body is examined by the pathologist is defenceless. The property and clothing can be handed to police officers as soon as it is removed and logged, just as is the case in a suspicious death. In this way no evidence is lost.

Upon completion of the autopsy the body should be examined by the forensic odontologist. In some cases the two may be able to examine the body together. The dentist should never precede the pathologist. Cleaning the teeth and face may remove trace or other evidence which is necessary to the investigation.

IDENTIFICATION

Much has been said by a variety of workers on this topic. It has to be completed expeditiously and the bodies returned to their relatives and friends without delay so that they may organise the appropriate rites. Undue delay only exacerbates grief and is unnecessary.

The process of making an identification is not complicated but it may not be easy. Essentially it consists of marrying together information received about people from their relatives, friends and others, with observations made in the mortuary. What is being done in this process is that an attempt is being made by the investigators to build up a picture of an individual, so that this could be shown to a stranger who would then be able to recognise that person from a photograph.

In practice what this means is that forms containing the information and observations are matched with one another. When an identity has been ascribed it is cross-checked with the remainder to make sure that there can be no confusion. It is at this stage that visual identification by relatives becomes practicable. In situations where there has been severe dismemberment this may not be advisable. All that can be done though is to advise against viewing the body. If the relatives are intent then they must be allowed to see the body.

DISCUSSION

Mass disasters are a complex series of events. Unravelling every sequence therein is not easy. It can only be done with any hope of success if the approach used is painstaking and systematic. Each shred of evidence has to be carefully preserved and examined and then placed into the context of the whole event. Whilst an isolated bit of information may have great evidential value, if it is taken in isolation its usefulness will not be fully appreciated. Pathologists deal in medical evidence but this on its own is only of limited value. It is also evidence which is easily corruptible, bodies have to be treated carefully and stored properly if this is not to happen. They must be put into refrigerated premises. Some commentators have suggested embalming before the bodies are examined. This is not acceptable under any circumstances because the nature of the process is such that there will inevitably be loss of evidence. Traces of explosives, for example, may be washed off, toxicological analyses would be difficult to interpret, as would histological studies. In the most extreme cases they would be impossible and so the whole investigation could be invalidated.

As the Mount Erebus disaster showed, deep freezing of bodies is the next best thing to refrigeration, thus in any situation where there may be delays before the bodies can be examined, this method of preservation should be used and not embalming. The latter should only be carried out after the post-mortem examination has been completed.

The pathologist is really looking at three groups of things, the injuries, disease processes and toxicological findings. Insofar as the injuries are concerned, the decision which has to be made initially is when did they occur? Did they occur before death, at the time of death or afterwards? Ante-mortem injuries may be entirely innocent; they usually are but they may indicate a prolonged accident sequence, thus an aeroplane may have been experiencing difficulties long before it crashed, and the occupants may have been thrown around the cabin, thus hurting themselves. Very occasionally they may indicate an attempt at murder. More than one murderer has tried to use a fire to mask the injuries they have caused. Post-mortem injuries may have been caused during removal of the body, especially if the wreckage is perilous. Wrong attribution of timing can lead to misinterpretation of causation and thus incorrect analysis.

The pathologist is also looking at injuries in another way. They must never be seen in isolation, even when there is only one victim. Patterns of injury are vitally important to reconstruction. Similar injuries may indicate similar modes of causation in different accidents. Dissimilar injuries in accidents which have a common sequence indicate some other factor is at work. Occasionally there may be an odd man out. Someone whose injuries are totally different from the rest. Such a finding is worthy of very detailed study for it could be that they hold the clue to the cause of the accident. This was the case in one of the Comet accidents. The aircraft disappeared near Rhodes. One man's injuries were different from the rest and he was shown to have been peppered with tiny fragments. Further analysis of these, combined with a close study of a cushion, lead to the conclusion that a bomb had exploded nearby this man's seat. This had caused the aircraft to break up at altitude.

In another accident, the finding of lower leg fractures in many people who had survived the impact and lived for some time and been seen to be alive by people looking at the wreck, but who died when the aeroplane eventually caught fire, was shown to be the result of poor seat design. Similar findings in other accidents have confirmed the value of this approach. It is in fact an everyday feature of medical practice.

Achieving these explanations of cause and effect in the context of accidents particularly is not something the pathologist does in isolation. Taking the example of aircraft accident investigation as a model again, definitive answers only come about when the whole accident sequence is considered. Thus the features of the flight are examined by an Operations Inspector. He is someone who has considerable flying experience and who can therefore interpret the actions of the pilot. The engineering aspects are analysed by an engineer who has been specially trained in the field of wreckage analysis.

The medical findings are discussed in the context of the observations made by these experts and any others, such as explosives experts and flight data recorder analysts. In this way they can be looked at as part of an accident sequence and not as an isolated event. It is very dangerous to do otherwise for false assumptions can easily be made, and if an incorrect conclusion is arrived at, then the wider considerations of future safety will not have been served. Equally the rightful demands of the injured and bereaved will not have been served.

Injury analysis is not easy, especially when large numbers of people have been killed and injured. In these cases special tools have to be used. One of these, which has in recent years achieved wider usage, is the system of injury scoring. There are many methods, none of which has been shown to be completely flawless. All of them consist of giving an injury a numerical score. In most cases the higher the score, the more severe the injury. Various statistical tools can then be used to look at the injuries.

It is convenient to divide the body up into various regions and to look at the incidence and severity of the different injuries to the various components of the area concerned. Thus the number of fatal lesions to the vault of the skull can be looked at. If the seating positions of the passengers is known, then the incidence of these in different parts of the aeroplane can be looked at. Moreover, the overall severity of injury of the whole accident and of the occupants of the different zones in an aircraft, bus, train or other disaster site can be examined. Special cases, such as loss of clothing due to accident forces or signs of an explosion can also be looked at. These sub sets can then be compared with the whole.

This approach proved particulary useful in analysing the data from the Air India accident in 1985. Here only 131 bodies out of a total of 329, which made analysis particularly difficult, moreover only a portion of the wreckage was recovered which added to the problems of the investigators.

By using the method of injury scoring and looking at the incidence and severity of regional injuries, it could be shown that it was a moderately severe accident when considered in terms of overall injury severity; some of the victims drowned. Other conclusions were reached by looking at other features of the accident, thus it was seen that there was a tendency for people sitting down the sides of the fuselage to lose their clothes. The severity of injury increased towards the rear and to the sides of the aircraft. It also became apparent that there were three phases of injury, at the original altitude, as the aeroplane came down and at impact with the sea, but there was no evidence to show that the emergency was anticipated. People at the sides of the aircraft showed signs of decompression and of flailing; that is they were twisted around having come out of the fuselage at altitude.

The absence of signs consistent with wearing lap belts indicated that the event was unanticipated. The pattern of clothing loss indicated that people had been thrown out of the aircraft at various altitudes. Moreover, the loss of clothing was greatest at the sides of the cabin and towards the rear.

These and other findings, such as evidence of a high vertical impact loading, particularly in people seated at the rear and in the middle section of the rear zone, all pointed to a mid-air break up but the absence of finite signs precluded any diagnosis of the possible cause of the break-up.

In any analysis of this kind a number of assumptions have to be made, the most important of which is that people are actually occupying their seats. Inevitably some will not be thus seated and this does detract from the analysis. Nevertheless past experience has shown that more passengers do occupy their alloted seat and the chances of everyone being out of their seat at the time of an accident are remote.

The fact that the wreckage was spread over the floor of the ocean and the trail extended for about five miles, making recovery extremely difficult, added to the problems in analysing this accident and in arriving at a proven cause.

Aircraft fires also present analytical problems because of the fact that fires consume much of the evidence, however, by using a comprehensive analytical approach, much that is of value may be discovered and useful reconstructions can be achieved. This was the case in the fire which involved the British Airtours Boeing 737 at Manchester, which also occurred in 1985. Here a combustion can in the left engine broke and part of it pierced an underwing inspection panel, thus releasing fuel onto the hot engine. As this happened during the take off run the pilot had to bring the aircraft to a halt. This was achieved quickly, without injuring anybody.

The smoke and flames quickly entered the fuselage and they
surrounded the rear part of the aeroplane, thus rendering the
two back doors and the left overwing exit inoperable. These
problems were compounded by difficulties in opening the right
front entrance and the right overwing exit. In the case of the
former, this was due to the fact that the chute became jammed
due to a minor design fault; which has since been corrected.
The overwing exit was difficult to get at because a seat arm
was partly in the way. It was also heavy to lift and two
people had to effect opening.

In the meantime people were desperately trying to get out.
Some were becoming overcome by toxic fumes. The action of
these was vividly described by many survivors who spoke of
choking fumes, feelings of sleepiness and blockage of the eyes,
amongst other symptoms.

It soon became obvious that many of those who died had done so
because of the inhalation of fire fumes. Toxicological
analyses showed that many had high levels of carboxyhaemoglobin
and cyanide. Some also had volatiles and other combustion
products in their blood. When the seating positions of these
people were looked at, it became clear that the levels found
tended to be higher in people who were seated towards the back
of the aircraft.

Later work on the electron microscopical findings of the lungs
of some of those who died in this fire and of people killed in
other fires has shown that minute particles of burning aircraft
are inhaled deep into the lungs. The exact significance of
this finding has not been fully evaluated yet. However it does
seem that it has some important connotations for safety
recommendations. Smoke hoods/masks have been suggested as a
way of improving survivability in aircraft fires, as have
sprinkler systems. These, though they will remove some of the
particles, are not guaranteed to remove them all and it seems
that both may have to be used if real benefits are to accrue.

Toxicological analyses are important in trying to establish
accident causes. In one accident the cockpit crew were
overcome by fumes from a defective heater. This prevented them
from flying correctly so they deviated from the normal flight
path and did not respond correctly to information from air
traffic control and eventually crashed into a mountain side
killing all of those on board. Similar defects were found in
other heaters. Thus although one tragedy had occurred others
were prevented.

Alcohol and flying do not go together. Tests done in America
have shown that experienced pilots begin to make serious
procedural errors with alcohol levels as low as 30 mgm% and
inexperienced pilots began to make then at 20 mgm%. Studies of
accidents show that there are almost twice as many accidents in
those who have taken alcohol in the first few minutes of a
flight than in those who have not. Also the same is true for
those in their first 100 hours of flying.

Disease is not a common cause of accidents. Most
incapacitation in aircrew results from gastroenterological
upsets. It is true though, that in the past there have been
accidents held to have been caused by such things as coronary
artery disease in aircrew. However, there have been cases of
other members of the cockpit crew taking over when one of them
had a heart attack. This is a standard part of aircrew
training. All airline crews have to show that they can take
over in an emergency. The most dramatic event of this nature
concerned a French aircraft, the captain of which suffered an
epileptiform attack just as they were lifting off. The
co-pilot successfully took over and there was no accident.

CONCLUSIONS

The aviation pathologists role is as a member of a team which
has been tasked with the investigation of an aircraft accident.
In the United Kingdom this is usually done by pathologists of
the Royal Air Force, who work with operations and engineering
inspectors from the Department of Transports Air Accidents
Investigation Branch. Their task is to look for any possible
medical cause for the accident, to analyse the medical findings
and as a result of this analysis to assist in the delineation.
Having done this, they then help to put forward safety
recommendations.

ACKNOWLEDGEMENT

I am grateful to the Director General, RAF Medical Services for
permission to publish this paper. The opinions expressed are
those of the author and in no way reflect those of MOD (Air).

LOCAL AUTHORITIES AND DISASTERS — THE POLITICAL AND PRESS RESPONSE

N. Grizzard, Disaster Prevention and Limitation Unit,
University of Bradford

The author of this paper was actively involved in the aftermath
of both the Bradford Fire and the Hillsborough Disaster. At
Bradford as a member of the Chief Executive's Office, he came
into the City Hall, on the Sunday morning after the fire and
worked as one of the Council's emergency team. By Monday
morning he had been appointed Fundraising Coordinator of the
Bradford Disaster Appeal, and this became his main work for the
next year.

After the Hillsborough Disaster, he was brought in by Liverpool
City Council to help with the Hillsborough Disaster Appeal. In
both cases he worked with a large number of Local Authority
staff, and it is on both his own and their experiences, that
this paper is based.

When Disaster strikes, it is usually not in office hours. Both
Bradford City and Hillsborough happened on Saturday afternoons,
a time when the traditional Local Authority Office Machine was
effectively shut.

However at both the Bradford Fire Disaster and Hillsborough a
number of civic leaders were present. At Bradford, when the
match was to celebrate the club's promotion to Division Two of
the Football League, A civic party together with the Council's
Chief Executive, and visitors from Bradford's twin towns were
present. At Hillsborough, because the match was an F.A. Cup
semi-final there were visiting civic dignatories.

Liverpool Council had established a Press and Public Relations
Unit, some months before the Disaster in April 1989. They

became Liverpool Council's lead staff immediately after Hillsborough. Working with the political leadership they handled a whole variety of issues, including the problems caused by articles in two national newspapers over allegations about the conduct of Liverpool fans at Hillsborough.

Staff from the Unit were seconded out to help the Social Services, who joined with the other Merseyside Local Authorities to mount a massive help-line. Other members went to work with the Hillsborough Disaster Appeal, where again there was great press involvement.

Bradford and Hillsborough were football tragedies, and in both cases the Local Authorities were able to provide help to the football clubs. Bradford City and Liverpool F.C., both shared a common problem, they were football clubs and did not have large press and media resources. They were both under great pressure from supporters and the creation of a link with the Local Authority, helped to a degree to ease that strain. Local Authorities possess large resource staff resources, compared to many private-sector organisations. When Disaster strikes, there is often a need to lend these resources, even if previously there has been little contact between the two bodies.

The Press Office, became a focus for both incoming and outgoing information, and the Press Officer assumed a very high profile role. The Press Officer's team was enlarged, and his manager and another more senior colleague, worked under his direct control. Everyone accepted that because it was an emergency situation there was a need to change priorities and work patterns

On the Political side, within hours of the Disaster, Senior Politicians agreed there would be no political in-fighting, and the Civic Response would come from the Lord Mayor. This small, but key decision, helped greatly in coping with the aftermath. The clear mandate given to those involved with the post-Disaster work, allowed activities to proceed swiftly.

164

For those working on the Press-side, the activities lasted
until the First anniversary, when the memorial was unveiled.
The first two weeks after the Disaster were the most intense
but there followed Press work over the Appeal Fund, the Public
Inquiry, and a myriad of inquiries that came through during the
year.

Four years later, after the Hillsborough tragedy, the Media
Officers of both Sheffield and Liverpool Council found
themselves in the fore-front. In Sheffield, they were contacted
on the afternoon by the Press , asking for a Council response.
As in the Bradford experience, they found that their role was
greatly enhanced and that they needed access to equipment such
as mobile phones, that were not standard issue.

Disaster Limitation - the coping with the event once it has
happened, must have as a main strand, the acknowledgement
within the Local Authority Emergency Plan, of the Press
Dimension. If the Press are involved so must the Political and
Civic Leadership for they along with their paid Management,
will be faced with the barrage of questions.

"What are you going to do about it?"

"Whose responsibility is it for safety?"

"When did you last carry out an inspection?"

"Wasn't there a nearly a similar accident in 1984?"

All are questions, which in the cold light of day, with time,
experts available, and no pressure being exerted, appear
harmless. Standing at the scene of a disaster it's a very
different matter.

After the Bradford Fire, it became obvious that one part of the
Emergency Plan that was lacking was a Media Response. As in all

crisis situations, actions had to be taken , and the response
was to enlarge the Press Office, to allow the Media, a good
flow of information. This early action was critical for the
Local Authority to successfully cope.

The presence of Bradford's Chief Executive at the Football
stadium, helped in many ways the Local Authority response.

In both Bradford and Hillsborough, the initial activity from
the Emergency Services moved from rescuing casualties into the
second phase of recovery of the dead and an assessment of the
situation. The position that Local Authority Managers both from
the paid and the elected leadership find themselves, is facing
the Press.

Facing the Press, is not an unusual or daunting experience,
when it's the political correspondent of either the local
evening or weekly paper. For most local authority managers,
both officers and councillors it 's a routine business,
especially as everone is usually on first name terms.

When it's the full team from the national media, and that means
the Daily's, the Sunday's, Radio the four home T.V. Channels,
and satellite channels - together with arc lights which
increase the room temperature dramatically, not forgetting the
foreign press, it's a different matter. There may be a 'pack'
of fifty people all crowding round and asking questions.

Disasters by their nature are media events, from the first
inklings that something has happened, through to the Public
Inquiry, the findings of the Inquiry, and even the first
anniversary. The amount of press coverage is huge, especially
in the days immediately after the disaster.

Conclusion and Recommendations

(1) All Local Authority Emergency Plans should ensure that the Press Role is given a high priority. Upto date communications equipment including mobile phones, a photo-copier ansd word-processor should be available and usable.

(2) The Political Leadership should have a plan for their response. In most , if not all Local Authorities, there is political disagreement, but after a Disaster, it is important there is a unified voice.

(3) Look at whether Press help can be given to the 'victim' of the Disaster.

For further reading a good reference is **Out of the Valley** published by the Chief Executive's Office of Bradford Council.

Nigel Grizzard is a member of the DPLU and a Management Consultant . He worked in the Chief Executive's Office of Bradford Council from 1976-1988 and was the Fundraiser for the Bradford Disaster Appeal. He helped set up the City's Burns and Plastic Surgery Research Unit in 1985, and is a founder of the DPLU.